BAND SPECTRA AND
MOLECULAR STRUCTURE

BAND SPECTRA AND MOLECULAR STRUCTURE

BY

R. DE L. KRONIG, PH.D.

UNIVERSITY OF GRONINGEN

CAMBRIDGE

AT THE UNIVERSITY PRESS

1930

CAMBRIDGE UNIVERSITY PRESS
Cambridge, New York, Melbourne, Madrid, Cape Town,
Singapore, São Paulo, Delhi, Tokyo, Mexico City

Cambridge University Press
The Edinburgh Building, Cambridge CB2 8RU, UK

Published in the United States of America by Cambridge University Press, New York

www.cambridge.org
Information on this title: www.cambridge.org/9780521292573

First published 1930
First paperback edition 2011

A catalogue record for this publication is available from the British Library

ISBN 978-0-521-07650-0 Hardback
ISBN 978-0-521-29257-3 Paperback

DEDICATED TO
MY PARENTS

PREFACE

This book is an elaboration of a series of lectures given by the author at Trinity College, Cambridge, in May 1929. Its preparation was guided by two considerations: that it could be read with profit by the experimental as well as the theoretical physicist, although intended primarily as a presentation of the theory, and that its size should not be excessive. Classifying the subject-matter under three headings: The statement of the theoretical results, their derivation, and their comparison with experiment, it was felt that the first ought to be as complete as possible. The reader hence will find that an effort has been made to enumerate fully those consequences of the theory which are rigorously established together with the restricting conditions under which they hold. Other results, as e.g. the theory of molecule formation, in which there are a great many hypothetical elements, have been given less space, and an appropriate warning has been attached to those features regarding which considerable doubt still exists.

In the derivation of the results the plan generally adhered to was to omit proofs which require the use of group theory. Besides, purely mathematical details, as e.g. in the evaluation of the integrals of Chap. III, determining the intensities of band lines, have often not been gone into. Full references have been given in these cases. Also an attempt has been made to write the book in such a way that the experimental physicist may omit certain sections, in which a knowledge of the methods of wave mechanics is essential, without the continuity of the train of physical ideas being thereby seriously interrupted. These sections have been indicated by asterisks at the beginning and at the end.

Finally, as far as the discussion of the experimental material in its bearing on the theory is concerned, the proposed size of the book restricted this to typical examples being given.

The bibliography at the end takes into account the literature up to about November 1929. In the theoretical part those papers were included in which new physical ideas were formulated for the first time as well as those in which quantitative results are established with the help of wave mechanics. Papers bringing only mathematical considerations on the basis of the old quantum theory were left out as being of purely historical interest. In the experimental part the rule applied was to include the important papers in which an analysis of bands actually is carried out, papers dealing simply with the determination of wave lengths as well as preliminary notices being excluded in order to keep the bibliography within reasonable bounds.

Most of the subject-matter of the book has been previously published in scientific journals. Among the new features attention may be called to the contents of Art. 2, in which the difference between the mathematical treatment of molecules with strong and weak coupling of the spin has been gone into with greater care than hitherto; to the extension in Art. 5 of the results of Fues to apply to a diatomic molecule with an angular momentum around the internuclear line; to the selection rule at the end of section 2, Art. 15; to the theoretical justification of an assumption made in the determination of Avogadro's constant from the coherent scattering of molecular gases in Art. 21; and to a correction term in the formula for the specific heat of NO in Art. 25. Also it is hoped that the terms "homonuclear" and "heteronuclear" for diatomic molecules with equal or unequal nuclei may find favour with the reader and come into general use. The symbols employed for the various quantum numbers are in agreement with the unified notation recently adopted by band spectroscopists.

I do not wish to let this opportunity pass without expressing my sincere thanks to the Syndics as well as to the Staff of the Cambridge University Press for the helpful way in which they have met all my wishes regarding the sometimes rather cumbersome notation and the arrangement of the text.

R. DE L. KRONIG

July 1930

CONTENTS

CHAPTER IV

MACROSCOPIC PROPERTIES OF MOLECULAR GASES

CHAPTER V

MOLECULE FORMATION AND CHEMICAL BINDING

INTRODUCTION

Through the fundamental researches of J. J. Thomson, Lenard, and Rutherford electrically charged particles, *electrons* and *atomic nuclei*, have been recognised as the units of which matter is built up. The electrons, all alike, carry a negative charge

$$e = 4 \cdot 77 \cdot 10^{-10} \text{ E.S.U.,}$$

the elementary charge, and possess a mass

$$m = 0 \cdot 899 \cdot 10^{-27} \text{g.} \quad \dots\dots\dots\dots\dots(1)$$

Moreover, in order to obtain an interpretation of a number of spectroscopic phenomena, especially of the multiplet structure and the anomalous Zeeman effect in line spectra, it has been found necessary to ascribe to them an internal angular momentum or spin of magnitude $\frac{1}{2}h/2\pi$, where

$$h = 6 \cdot 55 \cdot 10^{-27} \text{ erg sec.}$$

is Planck's constant, and a magnetic moment $eh/4\pi m$, i.e. one Bohr magneton[†]. The atomic nuclei carry positive charges, ranging from 1 to 92 times the elementary charge, and for any one of these nuclear charges there may exist nuclei with different masses, *isotopes*, as shown above all by the experiments of Aston on positive rays. The nuclear masses are approximately integral multiples of the mass of the hydrogen nucleus, the *proton*, with a mass

$$\mu_{\text{H}} = 1 \cdot 661 \cdot 10^{-24} \text{g.,} \quad \dots\dots\dots\dots\dots(2)$$

the small deviations from the integral relationship being most conveniently expressed by the packing fraction as defined by Aston. To the nuclei too it is in some cases necessary to ascribe a spin of magnitude $s_n h/2\pi$, where s_n is some integer or half-integer[‡],

[†] See e.g. Hund, *Linienspektren und periodisches System der Elemente*, Springer, 1927.

[‡] The index n signifies "nucleus." If referring to a particular element we shall replace n by the symbol of that element, thus writing, e.g., s_{H} in the case of the hydrogen nucleus.

while little more is known about the associated magnetic moment than that it generally seems to be much less than a Bohr magneton (see Art. 18). A closed system of electrons and nuclei is called an *atom* if there is only a single nucleus present; if it contains more than one nucleus we speak of a *molecule*; in particular of a *diatomic molecule* if there are just two nuclei, otherwise of a *polyatomic molecule*.

The spectra obtained from gases or vapours in emission when a suitable exciting agency, such as an electrical discharge, is applied, or in absorption when light with a continuous range of frequencies is passed through, have been classified according to their appearance into *line spectra* and *band spectra*. The former, characterised in general by the relatively small number of distinct frequencies occurring in them, must be ascribed to monatomic gases or vapours. The latter, which have received their name from the very large number of lines, lying often so close together as to give the appearance of a shaded band, have been recognised as due to molecules. In fact, they form the most important source of information in questions of molecular structure which we possess. Most of the experimental material in the realm of band spectra refers to diatomic molecules, and the theory too has consequently confined its attention almost exclusively to this type of molecule. The subsequent discussion will hence be especially concerned with diatomic molecules and gases.

The information about molecules derived from their band spectra throws much light on a number of *macroscopic properties of molecular gases*, such as their optical behaviour, their dielectric, magnetic, and thermal properties. Also the *theory of molecule formation* or *chemical binding* has profited considerably by the fuller understanding of molecular structure obtained from the spectral evidence. These topics will then form a natural continuation of a discussion of band spectra and at the same time furnish a valuable check on some of the conclusions arrived at.

CHAPTER I

THE ENERGY LEVELS OF DIATOMIC MOLECULES AND THEIR CLASSIFICATION BY MEANS OF QUANTUM NUMBERS

1. GENERAL FOUNDATIONS.

The measurable quantities in a band spectrum obtained from a gas in emission or absorption are the *wave-lengths* and the *intensities* of the individual band lines. Besides, the state of *polarisation* of the radiation may be investigated. In the theoretical discussion the *frequencies* ν of the lines, connected with their wave-lengths λ by the relation $\nu = c/\lambda$, where

$$c = 3 \cdot 10^{10} \text{ cm./sec.}$$

is the velocity of light *in vacuo*, are the more suitable quantities to work with. For, as was shown by Schwarzschild (81)[†], Heurlinger (25), and Lenz (51), these frequencies, just as the frequencies in atomic spectra, obey a *combination principle*. This means that for a given molecule there exists a sequence of values W_1, W_2, ... such that the frequency $\nu_{jj'}$ of any line in the spectrum may be represented as the difference of two of the W's, W_j and $W_{j'}$ say, according to the formula

$$h\nu_{jj'} = W_j - W_{j'}, \quad \ldots\ldots\ldots\ldots\ldots\ldots(1)$$

h being Planck's constant.

According to the atomic theory of Bohr a molecule, like any other closed atomic system, is capable of assuming any one of a discrete set of energy values W_1, W_2, ..., and the spectral frequencies of the molecule are to be found from the W's by means of equation (1), the *frequency condition*. The W's in that equation are thus to be regarded as the energies of the *stationary states* of the molecule under consideration.

The first step then in the interpretation of a band spectrum is

† The numbers (1), (2), ... refer to the bibliography at the end of the book.

on the empirical side to find the values W_1, W_2, ... from the frequencies ν, as obtained by measurements of wave-lengths in the laboratory. This procedure, the *term analysis*, forms a branch of experimental spectroscopy rather than spectral theory and will hence not be dealt with in this book. The theory, on the other hand, must furnish methods to calculate the W's if the number of electrons and the number and kind of atomic nuclei of which the molecule in question is composed are given.

The *wave mechanics* of de Broglie and Schrödinger† makes such calculations possible in principle. It starts from a partial differential equation, the *wave equation*, whose independent variables are the coordinates of the particles of which the dynamical system is composed, having the form

$$(\mathbf{H} - W)\Psi = 0, \quad \dots\dots\dots\dots\dots\dots\dots(2)$$

where $\mathbf{H}$ is a differential operator‡, obtained from the classical Hamiltonian function of the system, and W a constant. Those values of W for which there exists a solution Ψ, the *wave function*, finite and single-valued in the domain of all the independent variables§, are interpreted as the energy values of the system.

Although this prescription for finding the W's is quite definite, it is in practice generally not applicable on account of the great mathematical difficulties encountered in the solution of partial differential equations with many independent variables. Nevertheless it has been possible to make important advances in the interpretation of the stationary states of diatomic molecules by a semi-quantitative method which has already proved its usefulness in the case of atomic spectra and which may suitably be denoted as the *classification of energy levels by means of quantum numbers*.

† De Broglie, *Ann. de phys.* **2**, 22, 1925 ; Schrödinger, *Ann. d. Phys.* **79**, 361, 489, 734 ; **80**, 437 ; **81**, 109, 1926.

‡ Throughout the text operators are printed in Clarendon type, while vectors are denoted by Gothic letters.

§ In some cases the integrability of the electrical density, expressed in terms of Ψ, over the domain of the coordinates rather than the finiteness of Ψ must be postulated.

The physical significance of this procedure may be briefly outlined as follows:

On account of the fact that the nuclear masses according to equations (1) and (2) of the Introduction are several thousand times as great as the mass of an electron, while the forces acting on the two kinds of particles are of the same order of magnitude, the frequencies in the electronic motion will be very large compared to the frequencies characterising the nuclear motion. It is hence natural to study at first the electronic motion with the nuclei treated as fixed centres of force. The energy values of the various stationary states of which this system will be capable, corresponding to the various types of motion of the electrons, the *electronic levels*, will still depend upon the distance ρ between the two nuclei as a parameter. They will be distinguished by a set of quantum numbers, the *electronic quantum numbers*.

If the nuclei no longer are kept fixed but are allowed to move along a straight line so that ρ is variable, then, with the electrons in a given stationary state, the nuclei will tend to settle themselves at that distance apart for which the energy of the molecule has a minimum, provided such a minimum exists. But besides this equilibrium configuration of the nuclei there will be other stationary states of the nuclear motion corresponding to various degrees of vibration of the nuclei about their equilibrium position. With each stationary state of the electronic motion there is then associated a whole sequence of *vibrational levels*, specified by a *vibrational quantum number*.

When the nuclei are no longer confined to move along a straight line but are allowed any position in space, there will result for them, besides the possibility of vibration around their position of equilibrium, that of rotation about the centre of mass of the molecule. As a consequence there will be built up on each of the vibrational levels a set of *rotational levels*, distinguished by a *rotational quantum number*.

The method just described means mathematically a kind of

separation of the wave equation of a diatomic molecule in the co-ordinates characterising the electronic motion, the nuclear vibration, and the nuclear rotation. In the following article, where the procedure outlined will receive a rigorous justification, it will appear that the separation can be carried out only if certain small terms in the Hamiltonian operator of the system, representing an interaction between the three types of motion, are neglected. These neglected terms will later turn out to be of importance for an explanation of certain features in the fine structure of the spectrum and of the phenomena of perturbations and predissociation.

After the mathematical investigation of the wave mechanics of a diatomic molecule we shall return to a more detailed discussion of the electronic, vibrational, and rotational levels, their energies, and their characterisation with the help of quantum numbers.

2. WAVE MECHANICS OF DIATOMIC MOLECULES†.

*The methods for treating mathematically the motion of diatomic molecules in the quantum theory have been evolved by Slater (82), Born and Oppenheimer (6), and Kronig (48) in the case where the electrons are considered as point charges, and by Van Vleck (83) for electrons with a spin of magnitude $\frac{1}{2}h/2\pi$. Since this spin frequently plays an important rôle, we shall follow here closely the work of Van Vleck. The wave function Ψ of the diatomic molecule entering into the wave equation (2), Art. 1, depends then, besides on the *positional coordinates* of the particles of which the molecule is composed, on certain other coordinates, the *spin coordinates*, first introduced by Pauli‡.

Disregarding the translatory motion of the molecule as a whole,

† The asterisks at the beginning and end of this article and of subsequent parts of the text indicate, as mentioned in the preface, that the reader not familiar with the concepts of wave mechanics may omit these sections without much detriment to his understanding of the further development.

‡ Pauli, *Zeit. f. Phys.* **43**, 601, 1927.

which is of no interest in connection with its spectrum and its internal structure, we consider the centre of mass of the nuclei as being at rest at the origin of coordinates. To be precise we should rather consider the centre of mass of the nuclei and the electrons as being at rest, but on account of the smallness of the electronic mass as compared to that of the nuclei the error thus made is negligible. Besides a rectangular coordinate system xyz, fixed in space, we introduce a movable one $\xi\eta\zeta$, with the same origin, the positive ζ-axis of which is directed from nucleus 2 to nucleus 1, while the ξ-axis lies in the xy-plane with its positive direction so chosen that in turning from the z- to the ζ-axis through the angle θ between them, where $0 \leqq \theta \leqq \pi$, we are going in the positive sense around the ξ-axis. As positional coordinates we shall at first use the rectangular coordinates x_r, y_r, z_r of the electrons in the fixed coordinate system, the distance of separation ρ of the two nuclei, and the polar angles θ and ψ, specifying the orientation of the ζ-axis in the coordinate system xyz with z as the polar axis and ψ measured from the direction x in the positive sense around the z-axis. Ultimately we shall describe the system in terms of the rectangular coordinates ξ_r, η_r, ζ_r of the electrons in the movable coordinate system and the coordinates ρ, θ, ψ just defined.

Besides the positional coordinates enumerated above we must introduce spin coordinates, one for each electron, which shall specify the orientations of the spin vectors $\mathfrak{s}_r$ with respect to a given direction in space. We shall call them s_r when choosing as this direction the z-axis, and σ_r when using the ζ-axis as line of reference. Since in order to obtain an interpretation of the spectroscopic facts it is necessary to assume that the component of the spin of an electron in a given direction can only take two values, $+\frac{1}{2}$ and $-\frac{1}{2}$, measured in units $h/2\pi$†, the spin coordinates too are each capable of two values only, say a_r and b_r for s_r, α_r and β_r for σ_r, which signify parallel and antiparallel orientation of the corre-

---

† When referring to angular momenta in the following discussion we shall in general omit the factor $h/2\pi$.

sponding spin with reference to the z- and ζ-axes respectively. They are in this regard very different from the positional coordinates, which can take any one of a continuous range of values.

The wave function Ψ, which to begin with we write in the co-ordinates x_r, y_r, z_r, ρ, θ, ψ, s_r, we shall later express: (a) in the coordinates ξ_r, η_r, ζ_r, ρ, θ, ψ, σ_r, or (b) in the coordinates ξ_r, η_r, ζ_r, ρ, θ, ψ, s_r, depending on whether the influence of the electron spins on the motion of the molecule is large or small compared with that of nuclear rotation. It must then be denoted more accurately by

$$\Psi\ (x_r,\ y_r,\ z_r,\ \rho,\ \theta,\ \psi,\ s_r),$$

$$\Psi\ (\xi_r,\ \eta_r,\ \zeta_r,\ \rho,\ \theta,\ \psi,\ \sigma_r), \qquad \text{case } (a),$$

$$\Psi\ (\xi_r,\ \eta_r,\ \zeta_r,\ \rho,\ \theta,\ \psi,\ s_r), \qquad \text{case } (b).$$

Since each index s_r or σ_r may take two values, we can also say that for a molecule with f electrons we are dealing with an aggregate of 2^f functions Ψ depending only on the positional coordinates and hence of the type encountered in problems of wave mechanics where there is no spin.

After having decided upon the coordinates to be used for the description of the molecule, it becomes necessary to construct the *Hamiltonian operator* $\mathbf{H}$ entering in the wave equation (2), Art. 1. At present $\mathbf{H}$ has to be obtained from the classical Hamiltonian function H of the molecule according to prescriptions justified by experience. This Hamiltonian function is additively composed of two parts, the kinetic energy T of the particles and their energy of interaction V. If to begin with we describe the position of the electrons with the aid of their coordinates x_r, y_r, z_r in the fixed coordinate system and the position of the nuclei by ρ, θ, ψ, while the orientation of the spins is specified by means of the spin co-ordinates s_r with respect to the z-axis, then T is the sum of two parts, the kinetic energy of the electrons

$$T' = \frac{1}{2m} \sum_r (p^2{}_{x_r} + p^2{}_{y_r} + p^2{}_{z_r})$$

and that of the nuclei

$$T'' = \frac{1}{2\mu}\left(p_\rho{}^2 + \frac{1}{\rho^2}p_\theta{}^2 + \frac{1}{\rho^2\sin^2\theta}p_\psi{}^2\right).$$

m is the electronic mass, $\mu = \mu_1\mu_2/(\mu_1 + \mu_2)$ the reduced mass of the nuclei, and $p_{x_r}, p_{y_r}, p_{z_r}, p_\rho, p_\theta, p_\psi$ are the momenta conjugate to the corresponding coordinates. V we also write as the sum of two parts, V' and V''. As mentioned we shall later distinguish two cases, (a) and (b), for which it is convenient to make this subdivision in different ways. V_a' shall be the sum of all the terms arising from the electric and magnetic interactions of the electrons and nuclei excepting those due to the fact that the nuclei are moving so that V_a' is obtained from V by putting p_ρ, p_θ, p_ψ equal to zero in it. V_a' will thus in general be a function of all the positional coordinates $x_r, y_r, z_r, \rho, \theta, \psi$, of $p_{x_r}, p_{y_r}, p_{z_r}$, and of the spin vectors $\mathfrak{s}_r$ (since there is a magnetic moment of one Bohr magneton associated with each $\mathfrak{s}_r$, parallel to it), but not of p_ρ, p_θ, p_ψ. V_a'' is the sum of the interaction terms not contained in V_a'. V_b' is obtained from V by not only putting p_ρ, p_θ, p_ψ equal to zero in it but also $\mathfrak{s}_r$ so that the interaction energy of the spins is not contained in it. V_b'' then stands for those terms of V not present in V_b'.

According to wave mechanics we have to replace T' by the operator

$$\mathbf{T}' = -\frac{h^2}{8\pi^2 m}\sum_r\left(\frac{\partial^2}{\partial x_r{}^2} + \frac{\partial^2}{\partial y_r{}^2} + \frac{\partial^2}{\partial z_r{}^2}\right),$$

T'' by the operator

$$\mathbf{T}'' = -\frac{h^2}{8\pi^2\mu\rho^2}\left[\frac{\partial}{\partial\rho}\left(\rho^2\frac{\partial}{\partial\rho}\right) + \frac{1}{\sin\theta}\frac{\partial}{\partial\theta}\left(\sin\theta\frac{\partial}{\partial\theta}\right) + \frac{1}{\sin^2\theta}\frac{\partial^2}{\partial\psi^2}\right],$$
$$\ldots\ldots\ldots(1)$$

while in V we have to replace

$$p_{x_r}\text{ by }\frac{h}{2\pi i}\frac{\partial}{\partial x_r}, \quad p_{y_r}\text{ by }\frac{h}{2\pi i}\frac{\partial}{\partial y_r}, \quad p_{z_r}\text{ by }\frac{h}{2\pi i}\frac{\partial}{\partial z_r}.$$

Instead of p_ρ, p_θ, p_ψ in V'' we also have to introduce derivatives with respect to the corresponding coordinates, but since we shall neglect V'' later, we need not consider here the nature of this substitution. The quantities $\mathfrak{s}_r$, finally, occurring in V, are according to Pauli† also to be regarded as operators s_r whose scalar components $\mathsf{s}_{rx}, \mathsf{s}_{ry}, \mathsf{s}_{rz}$ in the coordinate system xyz have the property of giving rise to a linear transformation of the components $\Psi(s_1, \ldots s_r, \ldots)$ of the wave function Ψ, in which for brevity we have omitted the positional coordinates, $\mathsf{s}_{rx}\Psi$ signifying a new function Ψ' whose components $\Psi'(s_1', \ldots s_r', \ldots)$ are given by

$$\Psi'(s_1', \ldots s_r', \ldots) = \sum_{s_1, \ldots s_r, \ldots} \mathfrak{s}_{rx}(s_1', \ldots s_r', \ldots; s_1, \ldots s_r, \ldots)\, \Psi(s_1, \ldots s_r, \ldots)$$

with

$$\left.\begin{aligned}
\mathfrak{s}_{rx}(s_1', \ldots s_r', \ldots; s_1, \ldots s_r, \ldots) &= \tfrac{1}{2} \text{ if } s_t' = s_t, t \neq r; s_r' = a_r, s_r = b_r, \\
&= \tfrac{1}{2} \text{ if } s_t' = s_t, t \neq r; s_r' = b_r, s_r = a_r, \\
&= 0 \text{ otherwise,}
\end{aligned}\right\}$$

$$\ldots\ldots\ldots(2)$$

and each s_r in the summation taking its two values a_r and b_r. s_{rx} may thus be regarded as a matrix with the components (2). Similarly s_{ry} and s_{rz} may be regarded as matrices with the components

$$\left.\begin{aligned}
\mathfrak{s}_{ry}(s_1', \ldots s_r', \ldots; s_1, \ldots s_r, \ldots) &= -\tfrac{1}{2}i \text{ if } s_t' = s_t, t \neq r; s_r' = a_r, s_r = b_r, \\
&= \tfrac{1}{2}i \text{ if } s_t' = s_t, t \neq r; s_r' = b_r, s_r = a_r, \\
&= 0 \text{ otherwise,}
\end{aligned}\right\}$$

$$\ldots\ldots\ldots(3)$$

$$\left.\begin{aligned}
\mathfrak{s}_{rz}(s_1', \ldots s_r', \ldots; s_1, \ldots s_r, \ldots) &= \tfrac{1}{2} \text{ if } s_t' = s_t, t \neq r; s_r' = s_r = a_r, \\
&= -\tfrac{1}{2} \text{ if } s_t' = s_t, t \neq r; s_r' = s_r = b_r, \\
&= 0 \text{ otherwise.}
\end{aligned}\right\}$$

$$\ldots\ldots\ldots(4)$$

In case (a) we introduce into the wave equation the coordinates $\xi_r, \eta_r, \zeta_r, \rho, \theta, \psi, \sigma_r$, in case (b) the coordinates $\xi_r, \eta_r, \zeta_r, \rho, \theta, \psi, s_r$. For this purpose we have to make use of the relations connecting them with the coordinates $x_r, y_r, z_r, \rho, \theta, \psi, s_r$. We have in the first place

$$\left.\begin{aligned}
\xi_r &= -x_r \sin\psi + y_r \cos\psi, \\
\eta_r &= -x_r \cos\theta \cos\psi - y_r \cos\theta \sin\psi + z_r \sin\theta, \\
\zeta_r &= x_r \sin\theta \cos\psi + y_r \sin\theta \sin\psi + z_r \cos\theta.
\end{aligned}\right\}\ldots\ldots(5)$$

† Pauli, *Zeit. f. Phys.* **43**, 601, 1927.

From these equations there follows:

$$\left.\begin{aligned}
\frac{\partial}{\partial x_r} &= -\frac{\partial}{\partial \xi_r}\sin\psi - \frac{\partial}{\partial \eta_r}\cos\theta\cos\psi + \frac{\partial}{\partial \zeta_r}\sin\theta\cos\psi, \\
\frac{\partial}{\partial y_r} &= \frac{\partial}{\partial \xi_r}\cos\psi - \frac{\partial}{\partial \eta_r}\cos\theta\sin\psi + \frac{\partial}{\partial \zeta_r}\sin\theta\sin\psi, \\
\frac{\partial}{\partial z_r} &= \frac{\partial}{\partial \eta_r}\sin\theta + \frac{\partial}{\partial \zeta_r}\cos\theta.
\end{aligned}\right\} \quad \ldots(6)$$

Furthermore the functions $\Psi(\sigma_1, \ldots \sigma_r, \ldots)$ with the spins referred to the coordinate system $\xi\eta\zeta$ are related to the functions $\Psi(s_1, \ldots s_r, \ldots)$ with the spins referred to the coordinate system xyz according to Pauli and Van Vleck(83)† by the linear transformation

$$\Psi(\sigma_1, \ldots \sigma_r, \ldots) = \underset{s_1, \ldots s_r, \ldots}{\Sigma}\ S(\sigma_1, \ldots \sigma_r, \ldots; s_1, \ldots s_r, \ldots)\Psi(s_1, \ldots s_r, \ldots),$$

where

$$S(\sigma_1, \ldots \sigma_r, \ldots; s_1, \ldots s_r, \ldots) = S(\sigma_1; s_1) \ldots S(\sigma_r; s_r) \ldots (7)$$

with

$$\left.\begin{aligned}
S(\alpha_r; a_r) &= \cos\frac{\theta}{2}e^{\frac{i}{2}\left(\psi+\frac{\pi}{2}\right)}, & S(\alpha_r; b_r) &= i\sin\frac{\theta}{2}e^{-\frac{i}{2}\left(\psi+\frac{\pi}{2}\right)}, \\
S(\beta_r; a_r) &= i\sin\frac{\theta}{2}e^{\frac{i}{2}\left(\psi+\frac{\pi}{2}\right)}, & S(\beta_r; b_r) &= \cos\frac{\theta}{2}e^{-\frac{i}{2}\left(\psi+\frac{\pi}{2}\right)},
\end{aligned}\right\} \ldots(8)$$

each index s_r in the summation taking its two values a_r, b_r. We may hence write

$$\Psi(\sigma_1, \ldots \sigma_r, \ldots) = \mathbf{S}\,\Psi(s_1, \ldots s_r, \ldots), \quad \ldots\ldots(9)$$

regarding $\mathbf{S}$ as a matrix with the components (7).

In case (a), in order to obtain from the wave equation

$$[\mathbf{H}(x_r, y_r, z_r, \rho, \theta, \psi, \mathsf{s}_{rx}, \mathsf{s}_{ry}, \mathsf{s}_{rz}) - W]\Psi(x_r, y_r, z_r, \rho, \theta, \psi, s_r) = 0$$

the wave equation in the coordinates $\xi_r, \eta_r, \zeta_r, \rho, \theta, \psi, \sigma_r$, we proceed in two steps:

1. We introduce both in $\mathbf{H}$ and Ψ instead of $x_r, y_r, z_r, \rho, \theta, \psi$, the new positional coordinates $\xi_r, \eta_r, \zeta_r, \rho, \theta, \psi$ and find thus

$$[\mathbf{H}(\xi_r, \eta_r, \zeta_r, \rho, \theta, \psi, \mathsf{s}_{rx}, \mathsf{s}_{ry}, \mathsf{s}_{rz}) - W]\Psi(\xi_r, \eta_r, \zeta_r, \rho, \theta, \psi, s_r) = 0.$$

† We use the results of Van Vleck, who pointed out a slight error in the formulae of Pauli.

2. To refer the spins to the axes $\xi\eta\zeta$ instead of the axes xyz we introduce $\Psi(\sigma_1, \ldots \sigma_r, \ldots)$ by means of equation (9), i.e.

$$\Psi(s_1, \ldots s_r, \ldots) = \mathbf{S}^{-1}\Psi(\sigma_1, \ldots \sigma_r, \ldots).$$

In place of $\mathbf{S}_{rx}$, $\mathbf{S}_{ry}$, $\mathbf{S}_{rz}$ in $\mathbf{H}$ we must write now $\mathbf{S}_{r\xi}$, $\mathbf{S}_{r\eta}$, $\mathbf{S}_{r\zeta}$, which in the new coordinates are the same matrices as $\mathbf{S}_{rx}$, $\mathbf{S}_{ry}$, $\mathbf{S}_{rz}$ in the old. Multiplying the resulting equation by $\mathbf{S}$ in front we find

$$[\mathbf{SH}(\xi_r, \eta_r, \zeta_r, \rho, \theta, \psi, \mathbf{S}_{r\xi}, \mathbf{S}_{r\eta}, \mathbf{S}_{r\zeta})\,\mathbf{S}^{-1} - W]$$
$$\times \Psi(\xi_r, \eta_r, \zeta_r, \rho, \theta, \psi, \sigma_r) = 0$$

as the new wave equation.

In transforming the Hamiltonian operator let us first consider the part $\mathbf{T}' + \mathbf{V}_a'$, which in the new coordinates gives rise to

$$\mathbf{H}_a' = \mathbf{S}\,(\mathbf{T}' + \mathbf{V}_a')\,\mathbf{S}^{-1}.$$

Since the equations (5) and (6), of which use is made in step 1 outlined above, as well as the matrix $\mathbf{S}$ with the components (7) employed in step 2, are the same regardless of whether the coordinate system $\xi\eta\zeta$ is moving or whether it is fixed in a given position θ, ψ; $\mathbf{H}_a'$ will be identical with the Hamiltonian operator of the molecule when the nuclei are not movable but fixed in the position ρ, θ, ψ; $\mathbf{T}'$ being the kinetic energy of the electrons, $\mathbf{V}_a'$ according to definition the energy of interaction with fixed nuclei. $\mathbf{H}_a'$ will hence not contain θ and ψ nor derivatives with respect to ρ.

To transform $\mathbf{T}''$ we must remember that the differentiations with respect to θ and ψ occurring in equation (1) are to be carried out with x_r, y_r, z_r kept constant, and that in the relations connecting ξ_r, η_r, ζ_r and x_r, y_r, z_r the angles θ, ψ enter as well as in $\mathbf{S}$. Distinguishing differentiations with x_r, y_r, z_r kept constant from those with ξ_r, η_r, ζ_r kept constant by providing the differentiation sign ∂ with a prime in the former, we find with the aid of equations (5) in performing step 1 on $\partial'/\partial\theta$

$$\frac{\partial'}{\partial\theta} = \frac{\partial}{\partial\theta} + \sum_r \left(\frac{\partial}{\partial\xi_r}\frac{\partial'\xi_r}{\partial\theta} + \frac{\partial}{\partial\eta_r}\frac{\partial'\eta_r}{\partial\theta} + \frac{\partial}{\partial\zeta_r}\frac{\partial'\zeta_r}{\partial\theta} \right)$$
$$= \frac{\partial}{\partial\theta} + \sum_r \left(\zeta_r\frac{\partial}{\partial\eta_r} - \eta_r\frac{\partial}{\partial\zeta_r} \right).$$

Step 2 gives then

$$\mathbf{S}\frac{\partial'}{\partial\theta}\mathbf{S}^{-1} = \frac{\partial}{\partial\theta} + \sum_r\left(\zeta_r\frac{\partial}{\partial\eta_r} - \eta_r\frac{\partial}{\partial\zeta_r} - is_{r\xi}\right) = \frac{\partial}{\partial\theta} - i\,\mathfrak{M}_\xi,$$

where

$$\mathfrak{M}_\xi = \sum_r\left[-i\left(\eta_r\frac{\partial}{\partial\zeta_r} - \zeta_r\frac{\partial}{\partial\eta_r}\right) + s_{r\xi}\right]\ldots\ldots\ldots\ldots(10)$$

represents the ξ-component of angular momentum of the electrons, orbital as well as spin, in the coordinate system $\xi\eta\zeta$, measured in units $h/2\pi$. The result can be verified by substituting for $\mathbf{S}$ and $s_{r\xi}$ the matrices with the components (7) and (2) respectively. Similarly

$$\mathbf{S}\frac{\partial'}{\partial\psi}\mathbf{S}^{-1} = \frac{\partial}{\partial\psi} - i\,\mathfrak{M}_\eta\sin\theta - i\,\mathfrak{M}_\zeta\cos\theta,$$

where $\mathfrak{M}_\eta$ and $\mathfrak{M}_\zeta$ are defined analogously to $\mathfrak{M}_\xi$. We obtain thus finally the wave equation

$$\begin{aligned}
\Bigg\{&\mathbf{H}_a'(\xi_r,\,\eta_r,\,\zeta_r,\,\rho,\,s_{r\xi},\,s_{r\eta},\,s_{r\zeta})\\
&- B\left[\frac{\partial}{\partial\rho}\left(\rho^2\frac{\partial}{\partial\rho}\right) + \frac{1}{\sin\theta}\left(\frac{\partial}{\partial\theta} - i\,\mathfrak{M}_\xi\right)\sin\theta\left(\frac{\partial}{\partial\theta} - i\,\mathfrak{M}_\xi\right)\right.\\
&\left.+ \frac{1}{\sin^2\theta}\left(\frac{\partial}{\partial\psi} - i\,\mathfrak{M}_\eta\sin\theta - i\,\mathfrak{M}_\zeta\cos\theta\right)^2\right]\\
&+ \mathbf{S}\mathbf{V}_a''\mathbf{S}^{-1} - W\Bigg\}\,\Psi(\xi_r,\,\eta_r,\,\zeta_r,\,\rho,\,\theta,\,\psi,\,\sigma_r) = 0,\ \ldots\ldots\ldots(11)
\end{aligned}$$

with the abbreviation

$$B = \frac{h^2}{8\pi^2\mu\rho^2}.\ \ldots\ldots\ldots\ldots\ldots\ldots\ldots\ldots\ldots(12)$$

To find a solution of equation (11) we construct a function Ψ^0 which satisfies an equation differing by small terms from equation (11) and which has the form

$$\Psi^0(\xi_r,\,\eta_r,\,\zeta_r,\,\rho,\,\theta,\,\psi,\,\sigma_r) = \Phi(\xi_r,\,\eta_r,\,\zeta_r,\,\rho,\,\sigma_r)\,\mathrm{P}(\rho)\,\Theta(\theta,\,\psi).$$
$$\ldots\ldots\ldots(13)$$

The functions Φ, P, Θ must be finite in the domain of their variables, normalised, make Ψ^0 single-valued and obey the differential equations

$$[\mathbf{H}_a' - W_a'(\rho)] \Phi = 0, \quad \ldots\ldots\ldots\ldots\ldots(14)$$

$$\left[B \frac{\partial}{\partial \rho} \left(\rho^2 \frac{\partial}{\partial \rho} \right) - W_a'(\rho) - W_a''(\rho) - \overline{U}_a(\rho) + W^0 \right] P = 0, \ldots(15)$$

$$\left\{ B \left[\frac{1}{\sin \theta} \frac{\partial}{\partial \theta} \left(\sin \theta \frac{\partial}{\partial \theta} \right) + \frac{1}{\sin^2 \theta} \left(\frac{\partial}{\partial \psi} - i\Omega \cos \theta \right)^2 \right] + W_a''(\rho) \right\} \Theta = 0.$$
$$\ldots\ldots\ldots(16)$$

Here $W_a'(\rho)$ is such a function of ρ that equation (14) has a solution Φ fulfilling the above requirements. Ω is the value of the angular momentum of the electrons around the internuclear line in that state of the molecule with nuclei kept fixed in space which is characterised by the function Φ obtained from equation (14); for $\mathfrak{M}_\zeta$ is an integral of the motion of that system so that $\mathfrak{M}_\zeta \Phi = \Omega \Phi$, as will be shown in Art. 3, where we shall see also that Ω has one of the values $0, \pm 1, \pm 2, \ldots$ if the number of electrons is even, and one of the values $\pm \frac{1}{2}, \pm \frac{3}{2}, \pm \frac{5}{2}, \ldots$ if the number of electrons is odd. $W_a''(\rho)$ is another function of ρ so to be determined that equation (16) has a solution Θ obeying the requirements previously stated. $\overline{U}_a(\rho)$ is to be obtained from the operator

$$\mathbf{U}_a = B \left[(\mathfrak{M}_\xi^2 + \mathfrak{M}_\eta^2) - \frac{\rho^2}{\Phi} \frac{\partial^2 \Phi}{\partial \rho^2} - \frac{2\rho^2}{\Phi P} \frac{\partial \Phi}{\partial \rho} \frac{\partial P}{\partial \rho} - \frac{2\rho}{\Phi} \frac{\partial \Phi}{\partial \rho} \right]$$

$$+ B \left[i \mathfrak{M}_\xi \left(2 \frac{\partial}{\partial \theta} + \cot \theta \right) + \frac{2i}{\sin \theta} \mathfrak{M}_\eta \frac{\partial}{\partial \psi} \right.$$

$$\left. + \cot \theta \left(\mathfrak{M}_\eta \mathfrak{M}_\zeta + \mathfrak{M}_\zeta \mathfrak{M}_\eta \right) \right] + \mathbf{S} V_a'' \mathbf{S}^{-1} \ldots\ldots(17)$$

by taking the average over the electronic coordinates $\xi_r, \eta_r, \zeta_r, \sigma_r$:

$$\overline{U}_a = \sum_{\sigma_1, \ldots \sigma_r, \ldots} \int \Phi^* \mathbf{U}_a \Phi d\xi_1 d\eta_1 d\zeta_1 \ldots d\xi_r d\eta_r d\zeta_r \quad \ldots\ldots(18)$$

with the * denoting the conjugate. It can be proved that only the first two terms in the first bracket of $\mathbf{U}_a$ give a contribution to the average so that it may be computed without P being known as soon as Φ has been determined from equation (14). Since these terms do not contain the angles θ and ψ, $\overline{U}_a$ will be a function of ρ only. The constant W^0 in equation (15) finally must be so chosen that a finite solution P exists and plays the rôle of the energy value belonging to the approximate wave function Ψ^0. Equation (11) is obtained from the equation which Ψ^0 obeys by adding the terms $(\mathbf{U}_a - \overline{U}_a)\Psi^0$. This can be seen by letting the operators in equations (14), (15), (16) act on Ψ^0 instead of on Φ, P, Θ and subtracting the last two of the resulting equations from the first.

In case (b), where the wave equation is to be written in the coordinates ξ_r, η_r, ζ_r, ρ, θ, ψ, s_r, we proceed just as in case (a), except that in the transformation of $\mathbf{H}$ step 2 is omitted, since here we do not refer the spins to new axes. We thus get

$$\left\{ \mathbf{H}_b{}'(\xi_r, \eta_r, \zeta_r, \rho) - B\left[\frac{\partial}{\partial\rho}\left(\rho^2 \frac{\partial}{\partial\rho} \right) + \frac{1}{\sin\theta}\left(\frac{\partial}{\partial\theta} - i\mathcal{J}_\xi \right)\sin\theta\left(\frac{\partial}{\partial\theta} - i\mathcal{J}_\xi \right) \right.\right.$$
$$\left.\left. + \frac{1}{\sin^2\theta}\left(\frac{\partial}{\partial\psi} - i\mathcal{J}_\eta\sin\theta - i\mathcal{J}_\zeta\cos\theta \right)^2 \right] \right.$$
$$\left. + \mathbf{V}_b{}'' - W \right\}\Psi(\xi_r, \eta_r, \zeta_r, \rho, \theta, \psi, s_r) = 0. \quad\ldots\ldots\ldots(19)$$

Here $\mathbf{H}_b{}'$ is the Hamiltonian operator obtained by expressing $\mathbf{T}' + \mathbf{V}_b{}'$ in the new coordinates and belonging hence to the molecule with fixed nuclei and with the spins neglected, i.e. with the electrons treated as point charges, while

$$\mathcal{J}_\xi = \sum_r \left[-i\left(\eta_r \frac{\partial}{\partial\zeta_r} - \zeta_r \frac{\partial}{\partial\eta_r} \right) \right]$$

is the ξ-component of orbital angular momentum of the electrons in the coordinate system $\xi\eta\zeta$ with similar definitions for $\mathcal{J}_\eta$ and $\mathcal{J}_\zeta$.

As an approximate solution we construct here a function

$$\Psi^0(\xi_r, \eta_r, \zeta_r, \rho, \theta, \psi, s_r) = \Phi(\xi_r, \eta_r, \zeta_r, \rho, s_r)\, \mathrm{P}(\rho)\, \Theta(\theta, \psi) \ldots(20)$$

with functions Φ, P, Θ obeying the equations

$$[\mathbf{H}_b' - W_b'(\rho)]\,\Phi = 0, \quad \ldots\ldots\ldots\ldots\ldots(21)$$

$$\left[B\frac{\partial}{\partial\rho}\left(\rho^2\frac{\partial}{\partial\rho}\right) - W_b'(\rho) - W_b''(\rho) - \overline{U}_b(\rho) + W^0 \right] P = 0,\ldots(22)$$

$$\left\{ B\left[\frac{1}{\sin\theta}\frac{\partial}{\partial\theta}\left(\sin\theta\,\frac{\partial}{\partial\theta}\right) + \frac{1}{\sin^2\theta}\left(\frac{\partial}{\partial\psi} - i\Lambda\cos\theta\right)^2 \right] + W_b''(\rho) \right\}\Theta = 0.$$
$$\ldots\ldots\ldots(23)$$

The terms by which the wave equation obeyed by Ψ^0 differs from the real wave equation are now

$$(\mathbf{U}_b - \overline{U}_b)\,\Psi^0 \text{ instead of } (\mathbf{U}_a - \overline{U}_a)\,\Psi^0,$$

where

$$\mathbf{U}_b = B\left[(\mathcal{L}_\xi^{\,2} + \mathcal{L}_\eta^{\,2}) - \frac{\rho^2}{\Phi}\frac{\partial^2\Phi}{\partial\rho^2} - \frac{2\rho^2}{\Phi P}\frac{\partial\Phi}{\partial\rho}\frac{\partial P}{\partial\rho} - \frac{2\rho}{\Phi}\frac{\partial\Phi}{\partial\rho} \right]$$

$$+ B\left[i\mathcal{L}_\xi\left(2\frac{\partial}{\partial\theta} + \cot\theta \right) + \frac{2i}{\sin\theta}\mathcal{L}_\eta\frac{\partial}{\partial\psi} + \cot\theta\,(\mathcal{L}_\eta\mathcal{L}_\zeta + \mathcal{L}_\zeta\mathcal{L}_\eta) \right] + \mathbf{V}_b''.$$
$$\ldots\ldots\ldots(24)$$

Here use has been made of the fact, to be proved in Art. 3, that $\mathcal{L}_\zeta\Phi = \Lambda\Phi$, where Λ is an integer.

In case (a) the energy of interaction due to the spins is taken into account in building up Ψ^0, while in case (b) it is disregarded. The method of case (a) is therefore suitable when this interaction is large compared to the influence of nuclear rotation, while the method of case (b) is appropriate to a small interaction of the spins. If the two methods are used in this way, then Ψ^0 will represent an approximate solution, which for the time being we shall accept as sufficiently accurate. Equation (13) or (20), according to which Ψ^0 is the product of the three functions Φ, P, Θ, satisfying the equations (14), (15), (16) or (21), (22), (23), is the mathematical expression of the circumstance outlined in the previous article that the motion of the molecule may be considered as a super-position of electronic motion, nuclear vibration, and nuclear rotation. Later we shall consider the influence of the neglected terms.*

3. ELECTRONIC LEVELS.

As sketched in Art. 1 and as demonstrated rigorously in Art. 2 [see equations (14) and (21) there], the first step in investigating the energy levels W of a diatomic molecule is to consider the nuclei as fixed at a given distance ρ apart and to study the energy levels $W(\rho)$ of this simplified system, the *electronic levels*, which are still functions of the parameter ρ. If the influence of the electronic spins on the motion of the molecule is large compared to that of the nuclear rotation, we include it in the electronic motion [case (a) of the preceding article] and call the resulting energy levels $W_a'(\rho)$; otherwise we disregard it for the time being [case (b) of the preceding article] and denote the energy levels by $W_b'(\rho)$.

To find $W'(\rho)$ by integrating the wave equation (14) or (21), Art. 2, of the molecule with fixed nuclei directly will in general not be practicable on account of the large number of independent variables entering in these equations. In the case of the *hydrogen molecule ion*, H_2^+, with two singly charged nuclei and one electron, the lowest value $W'(\rho)$, the energy of the normal state, has been obtained by Burrau (8), who used a method of numerical integration to attack this quantum mechanical analogue of the classical problem of two fixed centra exerting Coulomb forces on a mass point, and by Guillemin and Zener (22), who used a perturbation method. The experimental evidence confirms the computations. [See Richardson and Davidson (110).]

In atoms, where similar difficulties are encountered, it has been found possible to characterise the stationary states by assigning quantum numbers to the individual electrons†. Thus a sodium atom in its normal state is said to contain two $1s$-electrons, two $2s$-electrons, six $2p$-electrons, and one $3s$-electron, so that the state may be designated by

$$(1s)^2 (2s)^2 (2p)^6 (3s).$$

† See e.g. Hund, *Linienspektren und periodisches System der Elemente*, Springer, 1927.

The possibility of such a classification in the case of atoms rests on the fact that the complicated interaction of the electrons may approximately be replaced by a suitable central shielding of the nuclear field, and that in the motion of an electron in such a shielded central, no longer Coulombian, field the dynamical variables can still be separated. In a molecule no such method of replacing the interaction of the electrons by a suitable shielding of the nuclei is known. Hence a satisfactory characterisation of the individual electrons by means of quantum numbers can at present not be carried out in the case of molecules.

To a given electronic configuration of the molecule as a whole there can, however, be ascribed several quantum numbers having a well-defined physical meaning as first pointed out by Hund (29). If to begin with we consider the energy of interaction by which the spins are coupled to the molecule reduced to zero, then the *component of the orbital angular momentum of the electrons along the line joining the nuclei* will be a constant of integration, this line being an axis of symmetry, and can hence, according to wave mechanics, have only one of the quantised values $\Lambda = 0, \pm 1, \pm 2, \ldots$. If $\Lambda \neq 0$, say $\Lambda = p$, where p is a positive integer, there will always be a state with $\Lambda = -p$ exactly coinciding with the first. For by reversing the sense of rotation of all the electrons around the line joining the nuclei, another quantum state of the molecule with the same energy but with the opposite angular momentum as the former will result. According to whether $|\Lambda| = 0, 1, 2, \ldots$ we call the electronic level a Σ-, Π-, Δ-, $\ldots$ level. The energy differences of the various electronic levels corresponding to different orbital configurations of the electrons will be of the same order of magnitude as in the case of atoms, i.e. the order of a few volts for the lower states.

If the spins of the electrons are taken into account, their effect can be described as follows. For a given orbital configuration of the electrons their spin vectors $\mathfrak{s}_r$ of magnitude $\frac{1}{2}$ are composed to form a *resultant spin* $\mathfrak{S}$ of magnitude S. If the number of electrons in the molecule is even, S is one of the numbers $0, 1, 2, ...$, if odd, one of the numbers $\frac{1}{2}, \frac{3}{2}, \frac{5}{2}, ...$ We call the electronic level a singlet-, doublet-, triplet-,... level according to whether $S = 0, \frac{1}{2}, 1, ...$, and indicate the *multiplicity* $2S + 1$ by an upper left-hand index; thus $^1\Sigma$ for a singlet-Σ-state, $^3\Delta$ for a triplet-Δ-state, etc.

The resultant spin S has a magnetic moment of $2S$ Bohr magnetons associated with it, which will be capable of interacting with the orbital motion of the electrons provided this gives rise to a magnetic moment. If the energy of interaction is sufficiently large so that we have to take the spins into account at this stage of the calculation, case (a), then the resultant spin is to be considered as coupled to the molecule, and the *component of the spin along the line joining the nuclei* takes the $2S + 1$ quantised values $\Sigma = -S, -S+1, ... S$. The *component of the total angular momentum of the electrons along the line joining the nuclei* takes hence the $2S + 1$ values $\Omega = \Lambda + \Sigma = \Lambda - S, \Lambda - S + 1, ... \Lambda + S$. For the same reasons as when the spin was neglected each state $\Omega = \Lambda + \Sigma$ will coincide with a state $-\Omega = -\Lambda - \Sigma$, so that for a given $|\Lambda|$ and S there will be in general $2S + 1$ component levels, distinguished by the value of $|\Omega|$, which we shall write as a lower right-hand index; thus $^2\Pi_{\frac{1}{2}}$ and $^2\Pi_{\frac{3}{2}}$ for the two levels of a $^2\Pi$-state with $|\Lambda| = 1$, $\Sigma = -\frac{1}{2}$ or $\frac{1}{2}$ and therefore $|\Omega| = \frac{1}{2}$ or $\frac{3}{2}$ respectively. The splitting will be of the same order of magnitude as the energy differences between the components of a multiple level in an atomic spectrum, and in general the component levels will lie equidistant. If the one with largest $|\Omega|$ has the greatest energy value, the multiple level is called *normal*, otherwise *inverted*. If the energy of interaction due to the spins is very small so that it is more appropriate to neglect it entirely for the time being, case (b), which is always true if $|\Lambda| = 0$, i.e. for Σ-states, since then there is no orbital magnetic moment, and sometimes also for $|\Lambda| \neq 0$, then the component of the spin along the internuclear line will not be quantised, and no splitting of the state will take place. Singlet-levels, for which the spin is zero, may evidently be considered as coming under both case (a) and case (b). We shall adopt the convention to consider them as case (b).

* The mathematical justification of the results just described must be obtained with the aid of equations (14) and (21), Art. 2. In case (a), $\mathbf{H}_a'$ is the Hamiltonian operator of the molecule with fixed nuclei expressed in the coordinates ξ_r, η_r, ζ_r, ρ and with the spins of the electrons referred to the $\xi\eta\zeta$-axes. If we neglect the spins at first and introduce cylindrical coordinates ρ_r, ϕ_r, ζ_r for each electron in place of ξ_r, η_r, ζ_r with the direction ζ as axis and ϕ_r measured from the $\xi\zeta$-plane, and if we further introduce instead of ϕ_2, ϕ_3, ... the relative coordinates $\chi_2 = \phi_2 - \phi_1$, $\chi_3 = \phi_3 - \phi_1$, ..., then $\mathbf{H}_a'$ will not contain ϕ_1 itself, but only derivatives with respect to ϕ_1. For in these coordinates changing ϕ_1 simply signifies a rotation of the molecule as a whole around the line joining the nuclei, and $\mathbf{H}_a'$ must be invariant against such a rotation since the internuclear line is an axis of symmetry. Equation (14), Art. 2, can then be satisfied by a function of the form

$$\Phi' = e^{i\Lambda\phi_1}\,\Phi''\,(\rho_r,\,\chi_2,\,\chi_3,\,...\,\zeta_r,\,\rho), \quad\ldots\ldots\ldots\ldots(1)$$

where Λ is an integer. The component of orbital angular momentum of the electrons along the line joining the nuclei $\mathfrak{L}_\zeta$ is given in the new coordinates by

$$\mathfrak{L}_\zeta = -\,i\frac{\partial}{\partial\phi_1} \quad\ldots\ldots\ldots\ldots\ldots\ldots(2)$$

so that

$$\mathfrak{L}_\zeta\,\Phi' = \Lambda\Phi'$$

according to equation (1).

If we now assume the presence of a spin $\mathfrak{s}_r$ for each electron, but if we still neglect the interaction energy due to these spins, then equation (14), Art. 2, can be satisfied by a function

$$\Phi = \Phi'\,\Xi\,(\sigma_1,\,...\,\sigma_r,\,...), \quad\ldots\ldots\ldots\ldots\ldots(3)$$

where Φ' is the function (1) and Ξ has the property of being unity for a certain combination of the spin coordinates, say e.g. for $\sigma_1 = \alpha_1$, $\sigma_2 = \beta_2$, ... $\sigma_f = \alpha_f$, and zero for all others. Since there are

2^f such combinations, where f is the number of electrons, there are 2^f functions Ξ and hence 2^f functions Φ, all belonging to the same energy value. Now not all of these wave functions will be realised in nature, for according to the exclusion principle of Pauli† only those states do actually occur, the wave functions of which are antisymmetrical in the electrons, i.e. the wave functions of which reverse sign if the coordinates of any pair of electrons are interchanged. If there were no spin, then only such states should occur for which the function Φ' of equation (1) itself is antisymmetrical in the positional coordinates of any two electrons. On account of the presence of the spin, however, it is frequently possible to construct antisymmetrical wave functions by linear combination of the Φ's of equation (3) even if Φ' of equation (1) is not antisymmetrical.

As an example the simplest case of two electrons may be mentioned. Let Ξ_{11} be the function equal to unity if $\sigma_1 = \alpha_1$, $\sigma_2 = \alpha_2$ and zero otherwise; Ξ_{12} the function equal to unity if $\sigma_1 = \alpha_1$, $\sigma_2 = \beta_2$ and zero otherwise, and similarly Ξ_{21}, Ξ_{22}. If Φ' is already antisymmetrical in the positional coordinates of the electrons, we can, by taking the spins into account, construct three antisymmetrical wave functions from it, viz.

$$\Phi'\,\Xi_{11}, \quad \Phi'\,\Xi_{22}, \quad \frac{1}{\sqrt{2}}\,\Phi'\,(\Xi_{12} + \Xi_{21}), \quad \dots\dots\dots(4)$$

while if Φ' is symmetrical in the positional coordinates, one antisymmetrical wave function,

$$\frac{1}{\sqrt{2}}\,\Phi'\,(\Xi_{12} - \Xi_{21}), \quad \dots\dots\dots\dots(5)$$

may be constructed from it. In the first case we have a triplet-state, in the second a singlet-state, and in general with any number of electrons it has been proved by Wigner and Witmer (85) with the aid of group theory that there results an odd multiplicity if this number is even, otherwise an even multiplicity. Just as in the

† Pauli, *Zeit. f. Phys.* **31**, 765, 1925; Dirac, *Proc. Roy. Soc.* **A 112**, 661, 1926.

case of two electrons the multiplicity will depend upon the symmetry properties of the function Φ' of the positional coordinates.

The component of angular momentum of the electrons along the line joining the nuclei $\mathfrak{M}_\zeta$ is in the presence of a spin no longer given by equation (2) but, according to equation (10), Art. 2, by

$$\mathfrak{M}_\zeta = -i\frac{\partial}{\partial\phi_1} + \sum_r s_{r\zeta}.$$

In the case of two electrons, if we let this operator act on the functions Φ given by the expressions (4) and (5), then, according to the definition (4), Art. 2, with $s_{r\zeta}$ written instead of s_{rz}, we find

$$\mathfrak{M}_\zeta \Phi' \Xi_{11} = (\Lambda + 1)\Phi' \Xi_{11}, \quad \mathfrak{M}_\zeta \Phi' \Xi_{22} = (\Lambda - 1)\Phi' \Xi_{22},$$
$$\mathfrak{M}_\zeta \frac{1}{\sqrt{2}} \Phi' (\Xi_{12} + \Xi_{21}) = \Lambda \frac{1}{\sqrt{2}} \Phi' (\Xi_{12} + \Xi_{21}), \quad \Bigg\} \quad (6)$$

$$\mathfrak{M}_\zeta \frac{1}{\sqrt{2}} \Phi' (\Xi_{12} - \Xi_{21}) = \Lambda \frac{1}{\sqrt{2}} \Phi' (\Xi_{12} - \Xi_{21}). \quad \ldots\ldots(7)$$

Equations (6) signify that the triplet-state consists of three component states with an angular momentum component along the line joining the nuclei $(\Lambda + \Sigma)$, Σ taking the values $1, 0, -1$, while according to equation (7) for the singlet-state Σ has only the one value 0. Quite generally it has been proved by Wigner and Witmer [85] that a state of multiplicity $2S + 1$ consists of the component states with an angular momentum $\Lambda + \Sigma$ about the internuclear line, where Σ takes the values $-S, -S+1, \ldots S$.

If the interaction energy by which the spins are coupled to the orbital motion and to each other is introduced, the component states just discussed will no longer coincide but will be separated. The components of orbital angular momentum and of spin along the line joining the nuclei will then no longer be rigorously constant individually, but their sum Ω does remain so, a result utilised in the derivations of Art. 2.

In case (b) we refer the spins to the axes xyz instead of $\xi\eta\zeta$, but we do not consider their interaction at all at this stage. With the Hamiltonian operator H_b' we show just as in case (a) that the component of orbital angular momentum along the internuclear line takes an integral value Λ.*

Although, as we have seen, it is not possible to assign quantum numbers to the individual electrons for a given stationary state of the molecule with fixed nuclei, a classification of the electronic levels in a more limited sense can be obtained which goes beyond their characterisation by means of the quantum numbers Λ, S, Σ, Ω just discussed. This classification is based on the circumstance, first investigated by Hund (30), that if the distance ρ between the nuclei is made infinite, two separate atoms will result, while by letting ρ approach zero the molecule is transformed into a single atom. In the first case the energy values of the electronic levels of the molecule will go over into the sum of the energy values of the resulting atomic levels; in the second case, after subtracting the mutual potential energy of the nuclei, into the energy values of an atom whose nuclear charge is equal to the sum of the charges of the originally separated nuclei. The first process can actually be carried out by increasing the vibrational energy of the nuclei until dissociation results, while the second is purely imaginary. The molecular electronic levels may then be specified more accurately by stating to which atomic levels they belong.

This *coordination of molecular to atomic levels*, which as we shall see in Art. 4 may in many cases be empirically determined, gives rise to the question: If two atoms in definite stationary states are brought close together or if in an atom in a definite stationary state the nucleus is imagined divided into two parts which then are separated from each other, what values have the quantum numbers Λ and S of the molecular electronic levels resulting by this process, how many are there of each kind, and in what sequence do these various levels lie?

The first and second parts of this problem have been considered by Hund (30), (31), (33), Mulliken (73), (75), (77), and Wigner and Witmer (85). We shall confine ourselves here to obtaining an answer with the help of the *vector model* used by the first two of these authors, while for the rigorous justification of the results with the aid of group theory the reader is referred to the paper of Wigner and Witmer.

The state of an atom may be characterised by the total orbital angular momentum L of its electrons and their resultant spin S. If $L = 0, 1, 2, \ldots$, we speak of an S-, P-, D-, $\ldots$ state and for $S = 0, \frac{1}{2}, 1, \ldots$ of a singlet-, doublet-, triplet-, $\ldots$ state indicated by the upper left-hand index $1, 2, 3, \ldots$; thus ^3S for a triplet-S-state. If we have two atoms with orbital angular momenta L_1 and L_2 and spins S_1 and S_2 respectively, then the effect of approaching them may be described as follows: Each of the vectors L_1 and L_2 has its component in the direction of the line joining the nuclei quantised, the components taking the values

$$\Lambda_1 = -L_1, -L_1 + 1, \ldots L_1; \quad \Lambda_2 = -L_2, -L_2 + 1, \ldots L_2$$

respectively. The values of the resultant component of orbital angular momentum of the molecule along the internuclear line $\Lambda = \Lambda_1 + \Lambda_2$ are obtained by combining each of the values Λ_1 with each of the values Λ_2. By carrying out these combinations we get for $|\Lambda|$ the following values

$$L_1 + L_2, \ L_1 + L_2 - 1, \ \ldots \ 0,$$
$$L_1 + L_2 - 1, \ \ldots \ 0,$$
$$\ldots\ldots\ldots\ldots\ldots\ldots$$
$$L_1 - L_2, \ \ldots \ 0,$$

where it has been assumed that $L_1 \geqq L_2$. The spins S_1 and S_2 are coupled to give the resultant spin S of the molecule according to

$$S = |S_1 - S_2|, \ |S_1 - S_2| + 1, \ \ldots S_1 + S_2.$$

Each of the values S combined with each of the values $|\Lambda|$ just found represents one molecular electronic level. As an example we may take $L_1 = 1, S_1 = \frac{1}{2}, L_2 = 1, S_2 = \frac{1}{2}$, i.e. both atoms in a ^2P-state.

We have now for $|\Lambda|$ the table

$$2, 1, 0,$$
$$1, 0,$$
$$0,$$

for S the two values 0 and 1. We get thus three $^1\Sigma$- and three $^3\Sigma$-states, two $^1\Pi$- and two $^3\Pi$-states, and one $^1\Delta$- and one $^3\Delta$-state. If the atoms are of the same kind, e.g. two N-atoms, but if they are in different stationary states j and j', then either atom 1 may be in state j and atom 2 in state j' or vice versa. For this reason the number of molecular electronic levels of the various kinds obtained by the method just described must be multiplied by a factor 2 in this special case.

The answer to the question as to the kind and number of molecular states resulting from a given atomic state specified by L and S if we consider the nucleus of the atom divided into two nuclei which are then separated from each other, is even simpler. The separation will cause the component of L along the internuclear line to be quantised so that it takes the values

$$\Lambda = -L, -L+1, \ldots L.$$

$|\Lambda|$ hence takes the values

$$L, L-1, \ldots 0,$$

while the spin S of the molecule is the same as that of the atom. Thus from a ^3P-state of the atom we get a $^3\Sigma$- and a $^3\Pi$-state of the molecule.

The energy values of the molecular electronic levels are functions of the nuclear separation ρ. After having determined the kind of molecular states and the number of each kind resulting from given atomic states in the two limiting cases of infinitely distant and of coinciding nuclei, it becomes important to know how the levels behave when ρ is varied. In order that a stable molecule may exist in one of these molecular states it is necessary that the energy of the state for a certain region of the parameter ρ should be less than the energy of the separate atoms, having a minimum somewhere in that region. Evidently the question in which we are interested here from the spectroscopic point of view is intimately related to the problem of molecule formation or chemical binding to be considered in the last chapter. There we shall see in special cases

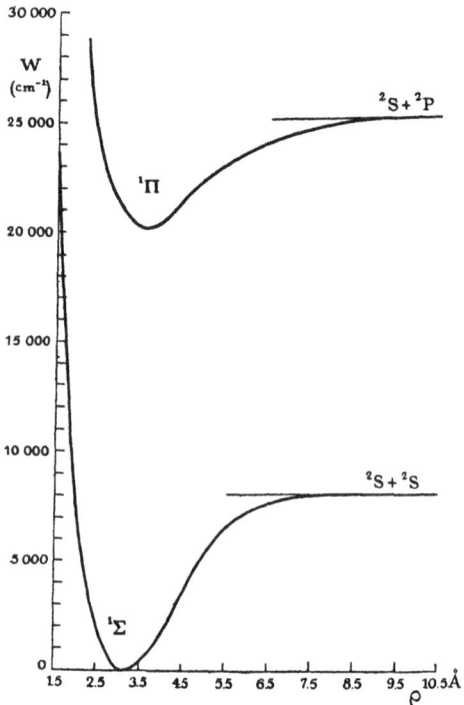

Fig. 1. Energy of two electronic states of Na_2 as function of internuclear distance.

that for some of the molecular states predicted by the vector model the above requirements are not fulfilled. These states are hence of no spectroscopic interest. For the states which do give rise to the formation of stable molecules Hund (33) and Mulliken (73), (75), (77) have tried to formulate semi-empirical rules giving the sequence in which their energies lie for various values of ρ. However, not being rigorous, these rules occasionally lead to predictions at variance with experiment so that care has to be exercised in their application. Thus the behaviour of the two $^2\Sigma$-states of N_2^+ as described by Hund is not in agreement with the results of Heitler and Herzberg (341), based directly on the empirical data.

To illustrate what has been said about the electronic levels of diatomic molecules in this article we show in Fig. 1 the energies of two electronic states of Na_2 as a function of the internuclear distance ρ, as given by Loomis and Nile (163). How these curves may be obtained from the experimental data we shall see in Art. 5. One state, the normal state of the molecule, is a $^1\Sigma$-level, the other one a $^1\Pi$-level. The integral value of S is in harmony with the even number of electrons in the molecule. The $^1\Sigma$-level gives for infinite ρ two Na-atoms in their normal state ^2S, the $^1\Pi$-level one normal Na-atom and one excited Na-atom in the lowest state ^2P. Accordingly the energy difference between the two asymptotes of the curves in Fig. 1 is equal to the energy difference between the ^2P- and the ^2S-state of Na.

4. VIBRATIONAL LEVELS.

The energy values $W'(\rho)$ of the electronic levels of the molecule with fixed nuclei in their dependence on ρ play essentially the rôle of the potential energy governing the vibration of the nuclei when these are allowed to move. As we have seen, in order that a stable molecule might exist at all in a given electronic state this potential energy must have a minimum for a certain value of ρ. If we still disregard in this article the possibility of rotation of the molecule as a whole, then we shall get from each electronic level a sequence of *vibrational levels* corresponding to various degrees of vibration of the nuclei around their position of equilibrium. We distinguish these by the *vibrational quantum number* v, to which we give the values 0, 1, 2, ... , the lowest vibrational level having $v = 0$.

For small values of v, i.e. for small amplitudes of vibration of the nuclei, their motion will be approximately simple harmonic and, as will be shown in Art. 5, the energy of the various states $v = 0, 1, 2, ...$

belonging to a given electronic level will be very nearly linear in v,

$$W = \text{const.} + c_1 v, \quad v = 0, 1, 2, \ldots, \quad \ldots\ldots\ldots\ldots(1)$$

where c_1 is a positive constant proportional to the vibrational frequency of the nuclei and may be directly determined from the spacing of the vibrational levels.

For larger values of v, i.e. for greater amplitudes of vibration of the nuclei, the restoring forces will get weaker on the average, as may be concluded directly from the type of curve shown in Fig. 1 if it be remembered that the mean distance between the nuclei will then become bigger. The vibrational frequency will hence decrease so that the vibrational levels come closer and closer until they finally *converge toward a limit* corresponding to the *dissociation of the molecule* into two separate atoms as first pointed out by Franck [19]. These atoms may be normal or excited, neutral or ionised, depending upon the particular electronic state under consideration. Indeed, if the energy value toward which the vibrational levels converge in a given case is known with sufficient accuracy, it is just from it that one may obtain information regarding the products resulting from the dissociation of the molecule when the energy of nuclear vibration is made large enough. Birge and Sponer [1] were able to show that the empirical data often are represented rather accurately by a formula of the type

$$W = \text{const.} + c_1 v - c_2 v^2, \quad v = 0, 1, 2, \ldots \ldots\ldots\ldots(2)$$

with positive constants c_1 and c_2. This formula gives levels coming closer together for increasing v and approaching the energy value

$$W_{\text{lim}} = \text{const.} + \frac{c_1^2}{4c_2}.$$

Sometimes it is necessary to add terms in v^3 and higher powers of v in equation (2) in order to obtain a satisfactory representation of the experimental data.

If in a diatomic molecule one of the nuclei has *isotopes*, the curves representing the energy values of the electronic levels as functions of ρ will be exactly identical for the molecules which may be built up with the various isotopes. For these curves refer to molecules with fixed nuclei whose distance of separation is varied as a parameter and in their construction therefore only the nuclear charge enters, which is the same for all isotopes, and not the nuclear mass. The position of the vibrational levels will, however, be different for the various kinds of molecules as shown by Kratzer (44) and Loomis (52). As long as the nuclear motion does not deviate too radically from simple harmonic vibration, the constant c_1 in equation (1), which is proportional to the vibrational frequency, will be inversely proportional to $\sqrt{\mu}$, where μ is the reduced mass of the nuclei, so that for the heavier isotope the spacing of the vibrational levels will be smaller.

The mathematical derivation of the various properties of the vibrational levels enumerated in this section we defer to the following article, where we shall consider the joint influence of vibration and rotation on the molecular motion on the basis of equations (15), (16) and (22), (23), Art. 2.

5. ROTATIONAL LEVELS.

When the diatomic molecule is allowed to rotate, two additional degrees of freedom are introduced into the motion besides those discussed in the last two articles. The effect of rotation is somewhat different, depending on whether the influence of the spin on the motion of the molecule is large or small compared to that of the nuclear rotation. In the first case the spin should be included in the electronic motion; we are dealing with case (a) of Arts. 2 and 3. In the second case we omit the interaction of the spin in the approximation considered here and have hence case (b) of Arts. 2 and 3.

In case (a) we have seen in Art. 3 that not only the component of orbital angular momentum Λ along the line joining the nuclei takes quantised values, but also the component of the spin Σ. If the molecule is now permitted to rotate, it will behave essentially like a symmetrical top with an angular momentum $\Omega = \Lambda + \Sigma$ around its axis of symmetry. On each of the vibrational levels belonging to the electronic state under consideration there will then be built up a set of *rotational levels*.

Since the *total angular momentum of the molecule* is a constant of integration in the absence of external fields, it will be quantised, and according to wave mechanics the square of its magnitude (measured in units $h/2\pi$) takes the values $J(J+1)$, where $J = |\Omega|$, $|\Omega| + 1, \ldots$. We can thus specify the rotational levels by means of the *rotational quantum number J*.

We wish now to find a formula which furnishes the energies of all the vibration-rotation levels v, J belonging to the various component levels Ω of one and the same electronic state q, Λ; q standing for the quantum numbers required besides Λ to specify the orbital electronic motion; e.g. of the levels $^2\Pi_{\frac{1}{2}}$ and $^2\Pi_{\frac{3}{2}}$ of the $^2\Pi$-state taken as an example in Art. 3. (If $S = 0$, there is just one component level; the electronic state is single.) This can be accomplished with the help of equations (15), (16), Art. 2, which govern the nuclear motion, and of which the first gives the energy levels W^0 associated with the approximate wave function Ψ^0 of equation (13), Art. 2. Following a procedure first introduced by Kratzer [43], we shall take the function $W_a{}'(\rho) + W_a{}''(\rho) + \overline{U}_a(\rho)$ in equation (15), Art. 2, to have the form

$$W_a{}'(\rho) + W_a{}''(\rho) + \overline{U}_a(\rho) = W^0{}_{q\Lambda\Omega}$$
$$+ (2\pi\nu_0)^2 \mu\rho_0{}^2 \left[\frac{1}{2} - \frac{\rho_0}{\rho} + \frac{1}{2}\left(\frac{\rho_0}{\rho}\right)^2 - c_3 \left(\frac{\rho - \rho_0}{\rho_0}\right)^3 - c_4 \left(\frac{\rho - \rho_0}{\rho_0}\right)^4 + \ldots \right]$$
$$+ B[J(J+1) + S(S+1) - 2\Omega\Sigma]\ldots\ldots\ldots(1)$$

for the electronic state under consideration. Here $W_{q\Lambda\Omega}$ is a quantity depending on the quantum numbers appearing as indices, while ν_0, ρ_0, c_3, c_4 depend upon q and Λ but not on Ω so that they are the same for all the multiplet components. B finally is the abbreviation (12), Art. 2. In particular c_3 and c_4 are supposed to be small. A justification of the above expression will be given in the mathematical section at the end of the article. The significance of ρ_0 and ν_0 may be seen as follows: Neglecting the term with B, which is small if J is not too large, we see that

$$W_a'(\rho) + W_a''(\rho) + \overline{U}_a(\rho),$$

which plays the rôle of the potential energy governing the nuclear vibration, has a minimum at $\rho = \rho_0$, taking there the value $W^0{}_{q\Lambda\Omega}$. ρ_0 may hence be regarded as the equilibrium distance between the nuclei. Still neglecting the term with B and developing the remainder in powers of $(\rho - \rho_0)$ we get from it

$$W^0{}_{q\Lambda\Omega} + (2\pi\nu_0)^2 \mu (\rho - \rho_0)^2 + \ldots.$$

ν_0 is thus the classical frequency of vibration of the nuclei for infinitely small displacements from their equilibrium position.

If we are satisfied with the approximate solution (13), Art. 2, of the wave equation (11), Art. 2, then, as will be discussed below, we get for the energy values W^0:

$$W^0{}_{q\Lambda\Omega v J} = W^0{}_{q\Lambda\Omega} - B_0 (\tfrac{7}{4} c_3 + \tfrac{7}{8} c_3{}^2 + \tfrac{9}{4} c_4)$$

$$+ h\nu_0 (v + \tfrac{1}{2}) \{1 - \tfrac{3}{2}\gamma^2 (1 + 2c_3) [(J + \tfrac{1}{2})^2 + S(S + 1) - 2\Omega\Sigma]\}$$

$$+ B_0 [(J + \tfrac{1}{2})^2 + S(S + 1) - 2\Omega\Sigma] \{1 - \gamma^2 [(J + \tfrac{1}{2})^2 + S(S + 1) - 2\Omega\Sigma]\}$$

$$- B_0 (v + \tfrac{1}{2})^2 (3 + 15c_3 + \tfrac{15}{2} c_3{}^2 + 3c_4) + \ldots \ldots (2)$$

with

$$v = 0, 1, 2, \ldots, \qquad J = |\Omega|, |\Omega| + 1, \ldots,$$

provided v and J do not take very large values. B_0 and γ are abbreviations

$$B_0 = \frac{h^2}{8\pi^2 \mu \rho_0{}^2}, \qquad \ldots \ldots \ldots \ldots (3)$$

$$\gamma = \frac{2B_0}{h\nu_0}, \qquad \ldots \ldots \ldots \ldots (4)$$

a quantity which in general is small.

If $W^0{}_{q\Lambda\Omega}$ is known, equation (2) furnishes the position of all the vibration-rotation levels v, J of the various component states Ω in the desired approximation. When the rotational frequencies become comparable to the frequency separation of the component levels of a multiple electronic state, equation (2) will no longer give good results. In the mathematical treatment the terms in the wave equation, neglected in finding the approximate wave function (13), Art. 2, to which the energy values (2) correspond, must then be taken into account, which will be done in Arts. 8, 9, and 10.

In case (b) the interaction of the spin is neglected in the molecule with fixed nuclei, and hence, as we saw in Art. 3, only the component of orbital angular momentum Λ along the internuclear line is quantised. If the molecule is then allowed to rotate, it will behave like a symmetrical top with an angular momentum Λ around its axis of symmetry. Now the *total angular momentum of the molecule, disregarding the spin,* will be quantised, its square taking the values $K(K+1)$, where $K = |\Lambda|, |\Lambda|+1, \ldots$, can here be called the *rotational quantum number.* Since now we do not take the spin into account in this approximation, then in order to obtain a formula for the energies of the vibration-rotation levels v, K belonging to a given electronic state q, Λ, we may simply take over equation (2) from case (a) if we only put S and Σ equal to zero and replace J by K. We find thus

$$W^0{}_{q\Lambda vK} = W^0{}_{q\Lambda} - B_0 \left(\tfrac{7}{4}c_3 + \tfrac{7}{8}c_3{}^2 + \tfrac{3}{4}c_4\right) + h\nu_0 \left(v + \tfrac{1}{2}\right)$$
$$\times \left[1 - \tfrac{3}{2}\gamma^2 (1 + 2c_3)\left(K + \tfrac{1}{2}\right)^2\right] + B_0 \left(K + \tfrac{1}{2}\right)^2 \left[1 - \gamma^2\left(K + \tfrac{1}{2}\right)^2\right]$$
$$- B_0 \left(v + \tfrac{1}{2}\right)^2 \left(3 + 15c_3 + \tfrac{15}{2}c_3{}^2 + 3c_4\right) \ldots(5)$$

with
$$v = 0, 1, 2, \ldots, \qquad K = |\Lambda|, |\Lambda|+1, \ldots.$$

Again we shall investigate in Arts. 8 and 10 the effect of the terms neglected in finding the approximate energy values (5).

If in equation (2) we put $J = |\Omega|$ or in equation (5) $K = |\Lambda|$, corresponding to vanishing nuclear rotation, we find for the energy of the various vibrational levels an expression having in first approximation the form (1), Art. 4, and in second approximation the form (2), Art. 4. Also for molecules built up of isotopes, as we shall see in the mathematical section of this article, the function obtained from $W_a'(\rho) + W_a''(\rho) + \overline{U}_a(\rho)$ of equation (1) by omitting the term with B should be practically identical so that $W^0_{q\Lambda\Omega}$, $\nu_0^2\mu$, ρ_0, c_3, c_4 are independent of μ, the reduced mass of the isotopic molecules. Therefore according to equation (2) for two such molecules with reduced masses μ' and μ'' respectively we have

$$\frac{B_0'}{B_0''} = \frac{\mu''}{\mu'}; \quad \dots\dots\dots\dots\dots\dots(6)$$

furthermore

$$\frac{\nu_0'}{\nu_0''} = \sqrt{\frac{\mu''}{\mu'}}, \quad \dots\dots\dots\dots\dots(7)$$

and according to equation (4) also

$$\frac{\gamma'}{\gamma''} = \sqrt{\frac{\mu''}{\mu'}}. \quad \dots\dots\dots\dots\dots(8)$$

Equations (6), (7), (8) allow us to find from equation (2) the position of the vibration-rotation levels belonging to the same electronic state in isotopic molecules. The same remarks apply in case (b).

If the vibration-rotation levels of an electronic state are known empirically, equations (2) and (5) will give us the constants ρ_0, ν_0, c_3, c_4. We can hence determine the shape of the curves

$$W'(\rho) + W''(\rho) + \overline{U}(\rho)$$

from a knowledge of the vibration-rotation levels. It is in this way that curves as shown in Fig. 1 may be obtained.

*A few words must still be said about the derivation of the energy values (2) and (5) from equations (15), (16) and (22), (23), Art. 2, respectively. In case (a) it will be quite safe to assume that in the neighbourhood of the equilibrium position of the nuclei the curves $W_a'(\rho)$ for the different multiplet components Ω of one and the same electronic state q, Λ run parallel so that we may write for all of them $W_a'(\rho) = W^0_{q\Lambda\Omega} + f(\rho)$, where f is independent of Ω. Nor will there be any difference in this expression for isotopic molecules since $W_a'(\rho)$ depends on the nuclear charge but not on the nuclear mass, having been obtained by keeping the nuclei of the molecule fixed in space. Furthermore equation (16), Art. 2, gives

$$W_a''(\rho) = B[J(J+1) - \Omega^2], \quad J = |\Omega|, |\Omega|+1,\dots,$$

as shown by Reiche [80] and by Kronig and Rabi [45], [46]. Of this we may incorporate the part $-B\Lambda^2$ in $f(\rho)$ without destroying its independence of Ω. As mentioned before, only the first two terms in the first bracket of expression (17), Art. 2, for $\mathbf{U}_a$ contribute anything to $\overline{U}_a(\rho)$. Remembering that $\mathfrak{M} = \mathfrak{L} + \mathfrak{H}$, where $\mathfrak{L}$ is the orbital, $\mathfrak{H}$ the spin angular momentum of the electrons, and that these two kinds of angular momentum are almost independent unless the multiplet intervals are very large, we find that $B\,(\overline{\mathfrak{M}_\xi^2} + \overline{\mathfrak{M}_\eta^2})$ is the sum of $B\,(\overline{\mathfrak{L}_\xi^2} + \overline{\mathfrak{L}_\eta^2})$ and $B\,(\overline{\mathfrak{H}_\xi^2} + \overline{\mathfrak{H}_\eta^2})$. The first is independent of Ω and may hence be incorporated in $f(\rho)$. $B\,(\overline{\mathfrak{H}_\xi^2} + \overline{\mathfrak{H}_\eta^2})$, according to quantum mechanics, is equal to $B[S(S+1) - \Sigma^2]$. The second term in $\mathbf{U}_a$ does not give anything depending essentially on Ω and may thus also be included in $f(\rho)$. We have thereby justified the assertion that ν_0, ρ_0, c_3, c_4 in equation (1) are the same for all multiplet components. We also asserted that $f(\rho)$ for isotopic molecules is the same. This is not quite accurate because the contributions from $W_a''(\rho)$ and $\overline{U}_a(\rho)$ to that function contain μ. But these contributions are themselves very small compared to the principal part of $f(\rho)$ coming from $W_a'(\rho)$ so that the error made is of a small order of magnitude.

The function (1) has to be substituted in equation (15), Art. 2, which shall give us the energy values W^0. For $S = \Sigma = 0$ the resulting equation has been treated by Fues [20], but his method is immediately applicable to our case†. The resulting energy values are those given by equation (2) as long as the development (1) is satisfactory, i.e. for vibrational and rotational quantum numbers not too large. In case (b) quite similar remarks apply to the derivation of the energy values (5).*

† The second term on the right-hand side of equations (2) and (5) is not correctly given by Fues, due to a mistake in the last step of his calculation.

6. Stark and Zeeman Effect.

When an *electric* or *magnetic field* is applied to a diatomic molecule, its energy levels are generally split into component levels. Just as in the case of atoms we speak of a *Stark* or *Zeeman effect*. Since the dynamics of a diatomic molecule is complicated as compared to that of an atom, the effect of the external fields will in general be much more involved in the former. For this reason too there is much less experimental material available for molecules than for atoms, and we shall therefore confine our attention in this article to those special cases where simple results may be expected.

For the purpose of investigating the influence of an electric field on the energy levels it is convenient to subdivide the diatomic molecules into two classes. If again we consider at first the nuclei as fixed, then the dipole moment of the molecule in a given electronic state may be zero on the average or it may contain a term independent of the time. In the latter case this average dipole moment must be parallel to the internuclear line for reasons of symmetry. Its magnitude we shall denote by $|\mathfrak{P}|$. In Art. 23 we shall see that $|\mathfrak{P}|$ vanishes always in *homonuclear molecules*, i.e. in molecules with two equal nuclei, while for *heteronuclear molecules* it will in general be different from zero.

In molecules without a constant dipole moment the only effect an electric field can produce is to deform the molecule. The result is a *quadratic Stark effect* of the same order of magnitude as that found in atomic spectra, e.g. in the case of Na, and hence exceedingly small in general. An example is the Stark effect in H_2 as studied by MacDonald (255). For molecules with a constant dipole moment we must again distinguish between electronic states which come under case (*a*) and those which come under case (*b*). In case (*a*) the *component of the total angular momentum in the direction of the electric field* will take the $2J + 1$ quantised values

$$M = -J, -J + 1, \ldots J.$$

As will be outlined in Art. 23, when considering the dielectric constant of molecular gases, the additional energy belonging to the various levels M due to an electric field of magnitude $|\mathfrak{E}|$ is in first approximation

$$W_{\mathrm{el}} = -\frac{\Omega M}{J(J+1)}|\mathfrak{P}||\mathfrak{E}|. \quad \ldots\ldots\ldots\ldots\ldots(1)$$

There results thus a *linear Stark effect*, provided Ω is different from zero. The higher approximations, proportional to $|\mathfrak{E}|^2$, $|\mathfrak{E}|^3$, etc., we shall not discuss here. In case (b) we have a similar result with Λ in place of Ω, K in place of J and M_K in place of M, where

$$M_K = -K, -K+1, \ldots K,$$

so that

$$W_{\mathrm{el}} = -\frac{\Lambda M_K}{K(K+1)}|\mathfrak{P}||\mathfrak{E}|.$$

For Σ-states, which always come under case (b) and for which $\Lambda = 0$, there is thus no linear Stark effect. As far as the author is aware, linear Stark effects have thus far not been obtained in molecular spectra.

In a magnetic field we are concerned with the magnetic moment of the molecule. In case (a) the molecule with fixed nuclei has a magnetic moment of $(\Lambda + 2\Sigma)$ Bohr magnetons parallel to its internuclear line, since for each unit of orbital angular momentum there is one Bohr magneton, while the ratio of magnetic moment to angular momentum must be taken twice as great for the spin in order to interpret the anomalous Zeeman effect in atomic spectra. In case (b) there is a magnetic moment of Λ Bohr magnetons parallel to the internuclear line and in addition the resultant spin S with a magnetic moment of $2S$ Bohr magnetons associated with it, which is not coupled to the rest of the molecule in the approximation considered.

In case (a) the *component of the total angular momentum along the direction of the magnetic field* of magnitude $|\mathfrak{H}|$ takes the $2J + 1$ quantised values

$$M = -J, -J+1, \ldots J,$$

and the additional energy of the corresponding component levels due to the field is given by

$$W_{\text{magn}} = \frac{\Lambda(\Lambda+2\Sigma)M}{J(J+1)}\frac{eh}{4\pi cm}|\mathfrak{H}|, \quad \ldots\ldots\ldots(2)$$

using a method sketched in Art. 24. There will thus result in general a *linear Zeeman effect*. In case (b) the total angular momentum K of the molecule, disregarding the spin, will have its component in the direction of the field quantised with the values

$$M_K = -K, -K+1, \ldots K,$$

and the same applies to the spin S, whose component takes the values

$$M_S = -S, -S+1, \ldots S.$$

The energy of a component level M_K, M_S is then given by

$$W_{\text{magn}} = \left[\frac{\Lambda^2 M_K}{K(K+1)} + 2M_S\right]\frac{eh}{4\pi cm}|\mathfrak{H}|. \quad \ldots\ldots\ldots(3)$$

It must be stressed here that for equations (2) and (3) to hold it is necessary that in case (a) the energy of interaction of the spin with the outside magnetic field be small compared to the energy of interaction with the orbital motion of the electrons, i.e. that the splitting produced by the magnetic field be small compared to the energy difference between the multiplet levels if the electronic state is multiple; also that it be small compared to the energy difference between successive rotational levels. In case (b) on the other hand the neglected interaction of the spin with the molecule must be small compared to the splitting caused by the outside field, while the second of the above requirements for case (a) must also be satisfied here. If these conditions are not fulfilled, complicated *Paschen-Back effects* will result. Singlet levels with $\Lambda \neq 0$ are hence best suited for obtaining simple Zeeman effects, and indeed the most satisfactory measurements on this phenomenon were made by Kemble, Mulliken, and Crawford (258) and by Crawford (256) in bands of CO which represent a transition between a $^1\Sigma$- and a $^1\Pi$-state. They give good agreement with the theory.

7. ENERGY LEVELS OF POLYATOMIC MOLECULES.

As shown by Born and Oppenheimer [6] it is possible in polyatomic molecules just as in diatomic molecules to consider the motion in first approximation as a *superposition of electronic motion, nuclear vibration, and nuclear rotation.* In the electronic motion, however, for an investigation of which the nuclei must again be regarded as fixed, on account of the absence of an axis of rotational symmetry there will now no longer exist so characteristic a quantum number as Λ in the case of diatomic molecules. Also the coordination of molecular to atomic levels when the nuclei are separated or made to coincide will become very much more involved. Since practically nothing is known experimentally about the electronic levels of polyatomic molecules, we shall not discuss them here any further.

The nuclear vibrations in a polyatomic molecule too will be much more complicated than in diatomic molecules. If the number of nuclei is $N > 2$, there will be $3N - 6$ vibrational degrees of freedom since the total number of degrees of freedom is $3N$, of which 3 are translational degrees of freedom of the centre of mass of the nuclei and 3 rotational degrees of freedom specifying the orientation of the nuclear structure in space. The various normal modes of vibration of such a system for small displacements of the nuclei from their positions of equilibrium have been investigated by Brester [7]. Dennison [12], [13] has applied his method to the molecules CO_2, NH_3, CH_4. Also from considerations on the polarisability of the ions of O, S, Se, and N, Hund [342], [343] has been able to conclude that the most stable configuration of the molecules H_2O, H_2S, and H_2Se is that in which the nuclei form an isosceles triangle with O, S, or Se at the vertex, while for NH_3 the H-nuclei form an equilateral triangle with the N-nucleus lying outside their plane and equidistant from them. His method permits also a rough calculation of the vibrational frequencies of these molecules. However, since only little is known experimentally regarding the nuclear vibrations of polyatomic molecules, we postpone a discussion of his work to the chapter on molecule formation.

When the nuclei are in their equilibrium position, the rotation of the molecule will resemble closely that of a rigid body with in general three different moments of inertia A_ξ, A_η, A_ζ. When two of the moments of inertia are equal, say $A_\xi = A_\eta$, then we have the problem of the symmetrical top. The energy values of this system in wave mechanics have been determined by Dennison (14), Reiche (80), and Kronig and Rabi (45), (46). They are found to be

$$W = \frac{h^2}{8\pi^2}\left[\frac{1}{A_\xi}J(J+1) + \left(\frac{1}{A_\zeta} - \frac{1}{A_\xi}\right)\Omega^2\right], \quad \ldots\ldots\ldots(1)$$

with $\qquad \Omega = 0, \pm 1, \pm 2, \ldots, \qquad J = |\Omega|, |\Omega|+1, \ldots.$

The general case of the rotation of a rigid body with three different moments of inertia has been investigated by Kramers and Ittmann (41), (42), by Wang (84), and by Ittmann (34). They do not find a closed expression for the energy values, but give those belonging to the lower quantum states. An experimental confirmation of equation (1) has been obtained by Badger and Cartwright (283), who studied the rotational levels of NH_3 in its normal state. A test of the results for a molecule with three different moments of inertia is very desirable.

CHAPTER II

FINE STRUCTURE AND WAVE MECHANICAL PROPERTIES OF THE ENERGY LEVELS OF DIATOMIC MOLECULES

8. The Perturbation Function.

In Chap. I we have derived approximate expressions for the energy levels of diatomic molecules as functions of various quantum numbers by neglecting certain terms in the wave equation, the influence of which we claimed to be small. For electronic states coming under case (a) these terms were denoted by $(\mathbf{U}_a - \overline{U}_a)\,\Psi^0$, for electronic states coming under case (b) by $(\mathbf{U}_b - \overline{U}_b)\,\Psi^0$, $\mathbf{U}_a$ and $\mathbf{U}_b$ being given by equations (17) and (24), Art. 2, respectively, and $\overline{U}_a$ and $\overline{U}_b$ signifying averages of $\mathbf{U}_a$ and $\mathbf{U}_b$ over the electronic coordinates as expressed for case (a) by equation (18), Art. 2, and for case (b) by a similar equation. It is the object of this and the following articles to investigate the influence of the neglected terms on the energy values of the molecule.

* The method appropriate to handle this problem is the *perturbation theory* of quantum mechanics†. Calling $(\mathbf{U}_a - \overline{U}_a)\,\Psi^0_{j'}$ or $(\mathbf{U}_b - \overline{U}_b)\,\Psi^0_{j'}$ for a given state j' of the molecule briefly $\mathbf{w}_a\,\Psi^0_{j'}$ or $\mathbf{w}_b\,\Psi^0_{j'}$, we develop this expression in terms of the unperturbed wave functions Ψ^0_j of all the states j of the molecule, thus

$$\mathbf{w}\,\Psi^0_{j'} = \sum_j w\,(j;\ j')\,\Psi^0_j,$$

the coefficients $w\,(j;\ j')$, the matrix elements of the perturbation energy, being given by

$$w\,(j;\ j') = \Sigma \int \Psi^0_j{}^* \,\mathbf{w}\,\Psi^0_{j'}\,d\tau, \quad\ldots\ldots\ldots\ldots\ldots(1)$$

where the integration extends over the domain of the positional coordinates, the summation over all values of the spin coordinates. The perturbation theory teaches how to express the energy values W_j

† Born, Heisenberg and Jordan, *Zeit. f. Phys.* **35**, 557, 1925.

of the perturbed system in terms of the energy values W^0_j of the unperturbed system and the quantities $w(j; j')$.

As regards the application of the perturbation theory to our problem, two important remarks have to be made. In the expression for the perturbed energy value of a state j the properties of all the other states j' as expressed by their energy values $W^0_{j'}$ and their wave functions $\Psi^0_{j'}$ enter. Now we have seen in Chap. I that in order to compute approximate energy values and wave functions for these various states, some are best considered as coming under case (a), others as coming under case (b), or in other words that for the first kind we neglect the part $\mathbf{w}_a\Psi^0$ of the wave equation, for the second kind the part $\mathbf{w}_b\Psi^0$. For the application of the perturbation theory, however, it is necessary to use the same perturbation function for all states, and for reasons which will become apparent later, we shall choose as such $\mathbf{w}_a\Psi^0$, in future simply denoted as $\mathbf{w}\Psi^0$, so that even those states of the molecule which approach closely the ideal case (b) are treated as case (a) in the initial approximation of the perturbation calculus.

Furthermore, it must be carefully considered if the unperturbed energy values have a degeneracy which is later removed by the perturbation in one of the approximations through which the computation is carried. States with $\Lambda \neq 0$, as we have seen in Art. 3, have such a degeneracy, two energy levels Λ, Σ and $-\Lambda$, $-\Sigma$ corresponding to the opposite senses of rotation of the electrons around the internuclear line coinciding. The same applies to the two levels Σ, $-\Sigma$ with $\Sigma \neq 0$ of a Σ-state, $\Lambda = 0$, which according to a previous remark is now also treated as case (a). Only the level $\Sigma = 0$ of a Σ-state is single. The degeneracy of spatial orientation we need not take into account since it is not removed by the perturbation function.

Let us now consider in order the various kinds of matrix elements $w(j; j')$. In the discussion purely mathematical details in connection with the evaluation of the integrals (1) will not be gone into, and the reader is referred for them to the paper by Van Vleck [83] already cited in Art. 2, who first applied the perturbation method to the most general diatomic molecule with spin. To distinguish the different states of the molecule it will be necessary to introduce again for j the whole set of quantum numbers q, Λ, Ω, v; J, M which we have used earlier.

There are in the first place the matrix elements

$$w\,(q,\ \Lambda,\ \Omega,\ v,\ J,\ M;\ q,\ \Lambda,\ \Omega,\ v',\ J',\ M'),$$

for which the quantum numbers q, Λ, Ω have the same values in the two states j and j', and which hence belong to two vibration-rotation levels of one and the same multiplet component Ω in a given electronic state q, Λ. Substituting for the functions Ψ^0 in equation (1) their expressions (13), Art. 2, the form of which will be discussed more specifically in Art. 12, and remembering the definition of $\overline{U}_a$ as given by equation (18), Art. 2, these matrix elements are all found to vanish, since for two such states the electronic parts Φ of the wave functions are identical.

For the following it will be advantageous to write $\mathbf{U}_a$ as the sum of three parts, $\mathbf{U}_a' + \mathbf{U}_a'' + \mathbf{U}_a'''$, where

$$\mathbf{U}_a' = B\left[(\mathfrak{M}_\xi^2 + \mathfrak{M}_\eta^2) - \frac{\rho^2}{\Phi}\frac{\partial^2\Phi}{\partial\rho^2} - \frac{2\rho^2}{\Phi\mathrm{P}}\frac{\partial\Phi}{\partial\rho}\frac{\partial\mathrm{P}}{\partial\rho} - \frac{2\rho}{\Phi}\frac{\partial\Phi}{\partial\rho}\right],$$

$$\mathbf{U}_a'' = B\left[i\,\mathfrak{M}_\xi\left(2\frac{\partial}{\partial\theta} + \cot\theta\right) + \frac{2i}{\sin\theta}\,\mathfrak{M}_\eta\frac{\partial}{\partial\psi} \right. $$
$$\left. + \cot\theta\,(\mathfrak{M}_\eta\mathfrak{M}_\zeta + \mathfrak{M}_\zeta\mathfrak{M}_\eta)\right],$$

$$\mathbf{U}_a''' = \mathbf{S}\mathbf{V}_a''\,\mathbf{S}^{-1}.$$

$(\mathbf{U}_a' - \overline{U}_a')$, $(\mathbf{U}_a'' - \overline{U}_a'')$, $(\mathbf{U}_a''' - \overline{U}_a''')$ will then be denoted by $\mathbf{w}'$, $\mathbf{w}''$, $\mathbf{w}'''$ respectively.

As explained by Van Vleck only those matrix elements $w'\,(j;j')$ will be different from zero for which the two states j and j' have the same quantum numbers Ω, i.e. those elements which are "diagonal" in Ω. Also since $\mathbf{U}_a'$ does not contain the angles θ and ψ, the elements $w'\,(j;\ j')$ in order not to vanish must be diagonal in J and M. Since the electronic part Φ of the wave function is independent of J and M, and the vibrational part P depends upon J only very slightly, the nuclear vibration being in general not greatly influenced by the nuclear rotation, and not at all on M, the same applies to the matrix elements $w'(j; j')$. We are hence only concerned with the following elements, from which the index M has been omitted:

$$w'(q, \Lambda, \Omega, v, J; q', \Lambda', \Omega, v', J), \quad (q', \Lambda') \neq (q, \Lambda)$$

and which are practically independent of J.

In evaluating the matrix elements $w''(j; j')$, we integrate in equation (1) at first only over the electronic positional coordinates ξ_r, η_r, ζ_r, and sum over the spin coordinates σ_r. As shown by Van Vleck, we get a result different from zero only if $\Omega' = \Omega \pm 1$, the quantities $\mathfrak{M}_\xi$, $\mathfrak{M}_\eta$, and $(\mathfrak{M}_\eta \mathfrak{M}_\zeta + \mathfrak{M}_\zeta \mathfrak{M}_\eta)$ becoming thereby the matrix elements

$$\mathfrak{M}_\xi(q, \Lambda, \Omega; q', \Lambda', \Omega \pm 1), \quad \mathfrak{M}_\eta(q, \Lambda, \Omega; q', \Lambda', \Omega \pm 1),$$

and

$$(2\Omega \pm 1)\, \mathfrak{M}_\eta(q, \Lambda, \Omega; q', \Lambda', \Omega \pm 1),$$

which still depend upon ρ as a parameter. From the relation

$$\mathfrak{M}_\xi(q, \Lambda, \Omega; q', \Lambda', \Omega \pm 1) = \mp i\mathfrak{M}_\eta(q, \Lambda, \Omega; q', \Lambda', \Omega \pm 1),$$

also mentioned in Van Vleck's paper, the result of the integrations in equation (1) thus far gone through is

$$2B\mathfrak{M}_\eta(q, \Lambda, \Omega; q', \Lambda', \Omega \pm 1) \left[\pm \frac{\partial}{\partial \theta} + \frac{i}{\sin \theta} \frac{\partial}{\partial \psi} + (\Omega \pm 1) \cot \theta \right].$$

Carrying out the integration over ρ and θ, ψ in equation (1) and using for the rotational part Θ of Ψ^0 the expression to be given in equation (7), Art. 12, one finds that only those elements $w''(j; j')$ are different from zero which are diagonal in J and M, and that these are given by

$$w''(q, \Lambda, \Omega, v, J; q', \Lambda', \Omega \pm 1, v', J)$$
$$= 2\,(BM_\eta)\,(q, \Lambda, \Omega, v, J; q', \Lambda', \Omega \pm 1, v', J)\, \sqrt{(J \mp \Omega)(J \pm \Omega + 1)},$$

$$\dots\dots(2)$$

the index M having been omitted since they do not depend on M.
Here

$$(B\mathfrak{M}_\eta)\, (q,\, \Lambda,\, \Omega,\, v,\, J;\ q',\, \Lambda',\, \Omega \pm 1,\, v',\, J)$$

$$= \int_0^\infty P^*{}_{q\Lambda\Omega vJ}\,[B\mathfrak{M}_\eta\,(q,\, \Lambda,\, \Omega;\ q',\, \Lambda',\, \Omega \pm 1)]\,P_{q'\Lambda'\Omega \pm 1 v'J}\,\rho^2 d\rho, \quad (3)$$

a quantity which will be practically independent of J since P, as mentioned before, varies in general only slightly with J.

The part $\mathbf{w}'''$ of the perturbation function, finally, we shall neglect. It arises from the magnetic fields due to the moving nuclei, and these will in general be small unless we go to high rotational quantum numbers. Moreover their effect will be partially compensated by that of the magnetic fields due to the additional motion which the electrons get from being carried around with the nuclei. Hence taking $\mathbf{w}'''$ into account would probably not modify the conclusions to be reached in the following articles as to the dependence upon the rotational quantum number of the changes caused in the energy values by the perturbation and might just alter the absolute magnitude of the shifts produced, which we shall not be able to evaluate accurately anyhow.

In the following articles we shall consider one by one the effects due to the various kinds of matrix elements which we have discussed. Again purely mathematical details will be omitted.*

9. ROTATIONAL DISTORTION OF SPIN MULTIPLETS.

By considering a diatomic molecule with the nuclei treated as fixed centres of force we have seen in Art. 3 that an electronic state q, Λ with $\Lambda \neq 0$, i.e. a Π-, Δ-, ... state, suffers a splitting into component levels distinguished by the quantum number Σ or Ω, provided $S \neq 0$, the number of these component levels being $2S + 1$. In Art. 5 we have shown how on each of the components there is built up a system of vibrational and rotational levels characterised by the quantum numbers v and J. If in equation (2), Art. 5, which expresses this result in mathematical form we give to v a definite value, we confine our attention to the rotational levels of the vibrational states with the same v belonging to the different multiplet components. Neglecting in equation (2), Art. 5, the small correction terms with γ^2 and calling the sum of $W^0_{q\Lambda\Omega}$ and those remaining terms which are independent of Ω and J, but two of which still depend upon v, $W^0_{q\Lambda\Omega v}$, we get for these levels

$$W^0_{q\Lambda\Omega vJ} = W^0_{q\Lambda\Omega v} + B_0 [(J + \tfrac{1}{2})^2 + S(S+1) - 2\Omega\Sigma]. \quad \ldots(1)$$

From equation (1) we see that the energy differences between the rotational levels with the same J belonging to the different multiplet levels do not vary with J.

The result just arrived at is not in agreement with experimental facts. In general the differences which we claimed to be constant diminish with increasing J. This phenomenon is the so-called *rotational distortion of spin multiplets*. An understanding of its cause is due to the qualitative work of Hund [29] and the computations on the basis of the older quantum theory performed by Kemble [36]. Later Hill and Van Vleck [26] and Van Vleck [83] have revised the theory using the concepts of wave mechanics.

* In order to interpret the distortion we can confine ourselves to taking into account those non-vanishing matrix elements of the perturbation function for which $q' = q$, $\Lambda' = \Lambda$, $v' = v$, and that are according to the preceding article only the elements

$$w'' (q, \Lambda, \Omega, v, J; q, \Lambda, \Omega \pm 1, v, J)$$

given by equation (2), Art. 8, which belong to two rotational levels with the same J in different multiplet components Ω and $\Omega \pm 1$ of one and the same electronic and vibrational state q, Λ, v. For them we have thus not only $\Omega' = \Omega \pm 1$ but more precisely $\Lambda' = \Lambda$, $\Sigma' = \Sigma \pm 1$, and we shall denote them hence for brevity by $w''_{q\Lambda vJ}(\Sigma, \Sigma \pm 1)$. Now $\mathfrak{M}_\eta$, the matrix elements of which occur in equation (3), Art. 8, and thereby also in equation (2), Art. 8, is the sum of $\mathfrak{L}_\eta$ and $\mathfrak{S}_\eta$, the η-components of the orbital and spin angular momenta of the electrons. Disregarding the slight interaction of the two kinds of angular momenta, which gives rise only to higher order effects, the components of $\mathcal{L}_\eta$ corresponding to $\Lambda' = \Lambda$ vanish so that $\mathfrak{M}_\eta$ may be put equal to $\mathfrak{H}_\eta$ for the matrix elements in which we are interested. But according to general quantum mechanical formulae the elements of $\mathfrak{H}_\eta$ corresponding to a uniform precession of $\mathfrak{S}$ around the internuclear line are given by

$$\mathfrak{S}_{\eta q\Lambda}(\Sigma;\ \Sigma \pm 1) = \tfrac{1}{2}\sqrt{S(S+1) - \Sigma(\Sigma \pm 1)},$$

independent of q and Λ, which specify the orbital motion of the electrons, and independent also of the internuclear distance ρ. In the integration over ρ in equation (3), Art. 8,

$$\mathfrak{M}_\eta\ (q, \Lambda, \Omega;\ q', \Lambda', \Omega \pm 1)$$

may hence be taken outside the integration sign, and the integral will become a sort of mean value of B, which will be obtained approximately from B by putting $\rho = \rho_0$ and is hence the quantity called B_0 in Art. 5. With the help of equation (2), Art. 8, we get then for the matrix elements of $\mathbf{w}$ with which we are concerned in this article

$$w''_{q\Lambda vJ}(\Sigma;\ \Sigma \pm 1)$$
$$= B_0 \sqrt{S(S+1) - \Sigma(\Sigma \pm 1)}\ \sqrt{J(J+1) - (\Lambda + \Sigma)(\Lambda + \Sigma \pm 1)},$$
$$\dots\dots(2)$$

remembering that $\Omega = \Lambda + \Sigma.$*

Since the degeneracies of the system present without the perturbation are not removed by it, the ordinary perturbation formulae may be applied, which state that the energy values of the perturbed system, up to and including second order terms, are given by

$$W_{q\Lambda\Omega vJ} = W^0_{q\Lambda\Omega vJ} + \frac{w''_{q\Lambda vJ}(\Sigma;\ \Sigma + 1)\, w''_{q\Lambda vJ}(\Sigma + 1;\ \Sigma)}{W^0_{q\Lambda\Omega vJ} - W^0_{q\Lambda\Omega + 1vJ}}$$
$$+ \frac{w''_{q\Lambda vJ}(\Sigma;\ \Sigma - 1)\, w''_{q\Lambda vJ}(\Sigma - 1;\ \Sigma)}{W^0_{q\Lambda\Omega vJ} - W^0_{q\Lambda\Omega - 1vJ}}.$$

Here $W^0_{q\Lambda\Omega vJ}$ is to be taken from equation (2), Art. 5. With the help of the expressions (2) the change produced in the energy values by the perturbation can be calculated for any multiplicity of the electronic level as long as the deviation from case (a) is not too large. Approximation formulae of this kind have been given by Hill and Van Vleck [26] for triplet states.

From equation (2) we see that the shift produced in the unperturbed energy values depends upon J only through a factor $J(J+1) + $ const. or, what is the same, $(J+\frac{1}{2})^2 + $ const. if we take for the unperturbed energy values the expression (1). The perturbed energy values depend then also upon J in the same way as the unperturbed ones according to equation (1), but with a proportionality constant different from that of equation (1) and different also for the various multiplet components. This situation may be described by saying that the molecule behaves as if it had an apparent moment of inertia depending on the component levels and different from the true moment of inertia.

For doublet states, as shown by Hill and Van Vleck [26] and by Van Vleck [83], it is possible to obtain an exact expression for the perturbed energy values, valid no matter how large the rotational distortion is. For then the states q, Λ, Σ, v, J perturb each other in pairs, viz. the states $q, \Lambda, \frac{1}{2}, v, J$ and $q, \Lambda, -\frac{1}{2}, v, J$. The perturbation theory teaches then that the perturbed energy values are the roots of the determinantal equation

$$\begin{vmatrix} W^0_{q\Lambda\frac{1}{2}vJ} - W & w''_{q\Lambda vJ}(-\frac{1}{2};\frac{1}{2}) \\ w''_{q\Lambda vJ}(\frac{1}{2};-\frac{1}{2}) & W^0_{q\Lambda-\frac{1}{2}vJ} - W \end{vmatrix} = 0,$$

where for convenience we have also labelled the energy values with the index Σ in place of the index Ω. Using for W^0 the expression (1), for the elements of $\mathbf{w}''$ the expression (2) with $S = \frac{1}{2}$, we find by solving this quadratic equation for W

$$W_{q\Lambda\pm\frac{1}{2}vJ} = W^0_{q\Lambda v} + B_0\left[(J+\tfrac{1}{2})^2 + \tfrac{1}{4}\right]$$
$$\pm \tfrac{1}{2}\sqrt{\Delta W^0_{q\Lambda v}(\Delta W^0_{q\Lambda v} - 4\Lambda B_0) + 4B_0^2(J+\tfrac{1}{2})^2}, \ldots\ldots(3)$$

provided $W^0_{q\Lambda\frac{1}{2}v} > W^0_{q\Lambda-\frac{1}{2}v}$ so that the doublet is normal. We have used the abbreviations

$$W^0_{q\Lambda v} = \tfrac{1}{2}(W^0_{q\Lambda\frac{1}{2}v} + W^0_{q\Lambda-\frac{1}{2}v}),$$
$$\Delta W^0_{q\Lambda v} = W^0_{q\Lambda\frac{1}{2}v} - W^0_{q\Lambda-\frac{1}{2}v}.$$

If $W^0_{q\Lambda\frac{1}{2}v} < W^0_{q\Lambda-\frac{1}{2}v}$ so that we are dealing with an inverted doublet, the signs in front of the square root have to be interchanged.

The correctness of the rules just stated can be tested by assuming that B_0 is small compared to $\Delta W^0_{q\Lambda v}$. Then for values of J not too large the square root may be developed in powers of $B_0/\Delta W^0_{q\Lambda v}$, and if we retain only terms up to and including the first power, we find again the unperturbed energy values (1) from which we started. In the other limiting case, B_0 very large compared to $\Delta W^0_{q\Lambda v}$ or J very large, we can develop the square root in powers of $\Delta W^0_{q\Lambda v}/B_0$. Retaining now only the first term of the development, we find the energy values

$$W_{q\Lambda+\frac{1}{2}vJ} = W^0_{q\Lambda v} + B_0 (J+1)^2,$$
$$W_{q\Lambda-\frac{1}{2}vJ} = W^0_{q\Lambda v} + B_0 J^2$$

if the doublet is normal, or with the left-hand sides of the equations interchanged if the doublet is inverted. Now the right-hand sides of these equations go over into the approximate energy expression (5), Art. 5, derived for a molecule coming under case (b), if in the upper one we put $J = K - \frac{1}{2}$, in the lower one $J = K + \frac{1}{2}$, while in equation (5), Art. 5, we neglect the terms with γ^2 and call the sum of those remaining terms which do not depend on K, $W^0_{q\Lambda v}$. We have thereby confirmed the applicability of formula (3) to describe the *transition from case* (a) *to case* (b), when the effect of nuclear rotation on the spins becomes large compared to that of the magnetic interaction by which they are coupled to the internuclear line. Also we see that for a normal doublet the levels $J, \Sigma = \frac{1}{2}$ and $J, \Sigma = -\frac{1}{2}$ go over respectively into the levels $J, K = J + \frac{1}{2}$ and $J, K = J - \frac{1}{2}$, while for an inverted doublet the coordination is just the reverse.

A very complete account of the experimental material in its bearing on the question of rotational distortion, particularly of $^2\Pi$-states, has been given by Mulliken (74). He finds a good agreement of the empirical data with equation (3) and the conclusions based on it regarding the different behaviour of normal and inverted doublets.

10. Fine Structure.

We have mentioned repeatedly that in a molecule with nuclei kept fixed in space a state having an orbital angular momentum Λ and a total angular momentum Ω along the internuclear line coincides with another state having a total angular momentum $-\Omega$ along that line. In the approximation considered in Chap. I, i.e. neglecting the perturbation terms in the wave equation discussed in Art. 8, this coincidence is not removed by the nuclear vibration and rotation. However, it is found empirically, at least for Π-states with $\Lambda = 1$, that every rotational level is actually resolved into two component levels, a phenomenon called σ-doubling in the earlier literature, where the component of total angular momentum along the internuclear line was usually denoted by σ, and now referred to as Λ-*doubling*. The energy difference between the component levels depends on the rotational quantum number J and, for multiple electronic states, on the multiplet component Σ. A theoretical interpretation of the experimental results has been given for singlet states by Hill and Van Vleck (26) and by Kronig (48), for doublet and triplet Π-states by Van Vleck (83).

According to the remarks just made the perturbation terms neglected in Chap. I must be held responsible for the phenomenon of Λ-doubling. In Art. 8 we have classified these terms, and in Art. 9 we have taken into account the influence of a part of them on the position of the energy levels, viz. of those which give rise to the rotational distortion of spin multiplets in multiple electronic states. The matrix elements to which this is due belong to pairs of rotational levels in different multiplet components of one and the same electronic state, $w''\,(q,\,\Lambda,\,\Omega,\,v,\,J;\ q,\,\Lambda,\,\Omega \pm 1,\,v,\,J)$. Now we must also take into account the matrix elements

$$w''\,(q,\,\Lambda,\,\Omega,\,v,\,J;\ q',\,\Lambda',\,\Omega \pm 1,\,v',\,J)$$

belonging to pairs of rotational levels in different electronic states $q,\,\Lambda$ and $q',\,\Lambda'$.

1. Λ-*doubling in singlet states*. Since for singlet-levels $\Omega = \Lambda$, we can denote the matrix elements in which we are interested here by w'' $(q, \Lambda, v, J; q', \Lambda \pm 1, v', J)$. They belong hence to electronic levels with Λ differing by one unit, thus for a $^1\Pi$-level, $\Lambda = 1$, to this $^1\Pi$-level and the $^1\Sigma$- or the $^1\Delta$-levels. The perturbation theory shows that these matrix elements lead in 2Λ-th approximation to a splitting of each originally single rotational level into two component levels with an energy difference having the form

$$\Delta W = \text{const.} \ (J - \Lambda + 1) \ (J - \Lambda + 2) \dots (J + \Lambda).$$

For $^1\Pi$-levels this becomes

$$\Delta W = \text{const.} \ J \ (J + 1), \quad \dots\dots\dots\dots\dots(1)$$

for $^1\Delta$-levels

$$\Delta W = \text{const.} \ (J - 1) \ J \ (J + 1) \ (J + 2).$$

The constant is of the order of magnitude

$$B_0 \left(\frac{B_0}{h\nu}\right)^{2\Lambda - 1}, \dots\dots\dots\dots\dots\dots(2)$$

where B_0 is given by equation (3), Art. 5, and ν is of the order of magnitude of an electronic frequency. B_0/h is of the order of magnitude of a rotational frequency and hence the factor in the parenthesis of the expression (2) is very small. From it we see that the Λ-doubling for $^1\Delta$-states will be much less than that for $^1\Pi$-states. Furthermore it follows that the doubling will be the greater the larger B_0, i.e. according to equation (3), Art. 5, the smaller the moment of inertia of the molecule. Experimentally it is found indeed that Λ-doubling can only be detected for $^1\Pi$-states, being already too small in $^1\Delta$-states, and that it is pronounced only in hydrides which, on account of the small mass of the hydrogen nucleus, have a small moment of inertia. An experimental confirmation of equation (1) for $^1\Pi$-states was first obtained by Bengtsson and Hulthén[195] in AlH-bands.

2. Λ-*doubling in* $^2\Pi$-*states.* Here we have to distinguish whether the $^2\Pi$-state comes under case (a) with the spin firmly coupled to the internuclear line and large separation of the multiplet

components, or whether we are dealing with case (b), where the spin is quite loosely coupled to the molecule, and the multiplet separation is negligible. In the latter case we evidently need not take the spin at all into account, and we shall have the same result as in the previous section with K in place of J, i.e. a doublet interval

$$\Delta W = \text{const.} \, K \, (K + 1).$$

In case (a) a rather involved perturbation calculus with the matrix elements responsible for Λ-doubling previously mentioned gives for $^2\Pi_{\frac{1}{2}}$

$$\Delta W = \text{const.} \, (J + \tfrac{1}{2})$$

with a constant of the order of magnitude

$$B_0 \frac{\Delta \nu}{\nu},$$

where B_0 is the expression (3), Art. 5, $\Delta \nu$ the spin doublet width of the $^2\Pi$-state, and ν of the order of magnitude of an electronic frequency. For $^2\Pi_{\frac{3}{2}}$ we get

$$\Delta W = \text{const.} \, (J - \tfrac{1}{2}) \, (J + \tfrac{1}{2}) \, (J + \tfrac{3}{2})$$

with a constant of the order of magnitude

$$B_0 \cdot \frac{B_0}{h \Delta \nu} \cdot \frac{B_0}{h \nu},$$

the quantities B_0, $\Delta \nu$, ν having the same significance as before. Examples of Λ-doubling for $^2\Pi$-states are fully discussed by Mulliken[76], who in general finds satisfactory agreement with experiment.

3. Λ-*doubling in* $^3\Pi$-*states.* Here again we must distinguish case (a) and case (b). In the latter the result is the same as for $^2\Pi$-states. In the former the splitting produced differs again, depending upon which of the three levels $^3\Pi_0$, $^3\Pi_1$, $^3\Pi_2$ we are considering. For $^3\Pi_0$ the energy difference of the two component levels is given by

$$\Delta W = \text{const.}$$

independent of J. It has the order of magnitude

$$\Delta \nu \cdot \frac{\Delta \nu}{\nu},$$

where $\Delta\nu$ is the frequency width of the triplet, ν of the order of an electronic frequency. For $^2\Pi_1$

$$\Delta W = \text{const. } J (J + 1),$$

the constant being of the order of magnitude

$$B_0 \cdot \frac{B_0}{h\nu}.$$

Finally for $^3\Pi_2$ the splitting is negligibly small. Again the reader is referred to the paper of Mulliken (76) for a comparison of the rather copious experimental material with the theoretical predictions, which it in general confirms as well as may be expected from the accuracy of the perturbation calculation.

Thus far we have only discussed the fine structure of electronic states with $\Lambda \neq 0$, i.e. of Π-, Δ-, ... states. Σ-states too have a fine structure if the spin $S \neq 0$. For a Σ-state, as explained in Art. 3, always comes under case (b), the spin being coupled very weakly to the molecule and hence not considered at all in the approximation of Chap. I. In the rotating molecule the perturbation terms, which also gave rise to Λ-doubling in Π-, Δ-, ... states, cause a rotational level corresponding to an angular momentum K of the molecule, disregarding its spin, to be split into component levels described by a quantum number J with the values

$$J = |K - S|, |K - S| + 1, \dots K + S.$$

This phenomenon is usually called ρ-"*doubling*," although, except for $S = \frac{1}{2}$, there are more than two component levels. We may say that the orbital angular momentum K and the spin S are combined to give the resultant J. For $^2\Sigma$-states with $S = \frac{1}{2}$ we hence expect each rotational level K to be split into two component levels $J = K \pm \frac{1}{2}$, excepting the level $K = 0$ which remains single with a value $J = \frac{1}{2}$. For the doublet interval the perturbation theory gives, according to Van Vleck (83),

$$\Delta W = \text{const. } (K + \frac{1}{2}).$$

For $^3\Sigma$-states three instead of two fine structure levels are to be expected with $J = K - 1$, K, $K + 1$ except for $K = 0$, from which we get a single level $J = 1$ only. The interpretation of the energy differences of the component levels makes it necessary to take into account the interaction of the spins of the individual electrons and has been worked out by Kramers[39], [40] with the help of group theory.

11. PERTURBATIONS AND PREDISSOCIATION.

We have seen in Art. 8 that certain pairs of energy levels j and j' of a diatomic molecule perturb each other, i.e. that the matrix elements $w(j; j')$ of the perturbation function are different from zero for them. Due to this mutual influence the true energy values of the levels will be displaced with respect to the unperturbed energy values. The perturbation theory teaches that the displacement depends upon two factors, the magnitude of $w(j; j')$ and the difference in energy of the unperturbed levels. The greater $w(j; j')$ and the closer together the perturbing levels, the greater will be the change in energy produced.

We have seen in Art. 8 that only energy levels with equal J are able to perturb each other. Considering the rotational levels J of a given multiplet component of a certain electronic state, they can be perturbed by the rotational levels with equal J belonging to the remaining multiplet components of the same electronic state or by the rotational levels with equal J belonging to different electronic states. The first effect is that studied in Art. 9, the rotational distortion of spin multiplets. It shifts the various levels J by amounts varying regularly with J so that if their energies are plotted as a function of J, a smooth curve still results. The effect due to the rotational levels of another electronic state also changes regularly with J if the electronic state lies far away from the given one. But it may happen occasionally that the rotational levels of the two electronic states lie as indicated in Fig. 2. Then it is seen that the energy differences of the pairs of levels which perturb each other and which are connected by dotted lines vary rapidly with J, changing sign at a certain place. Since the change in energy brought about by the perturbation depends upon these energy differences, as mentioned above, being greater the smaller the difference, there will result an irregularity in the position of the perturbed energy values. Instead of lying on the smooth curve of Fig. 3 they may be expected to lie as indicated by the circles,

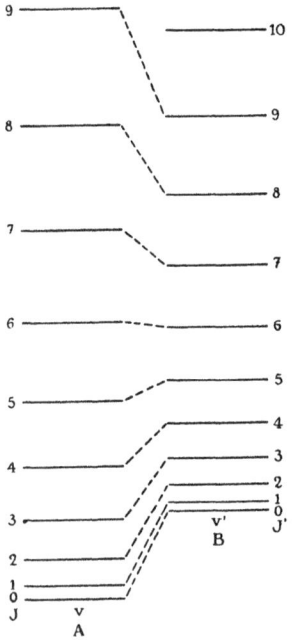

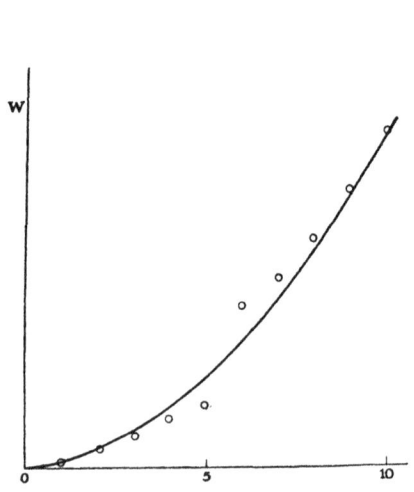

Fig. 3. Energy of rotational levels belonging to the state A, v of Fig. 2 as function of J.

Fig. 2. Rotational levels J and J' of given vibrational levels v and v' in two different electronic states A and B of a diatomic molecule. The levels connected by dotted lines can perturb each other.

the place at which their position changes from below the curve to above the curve corresponding to the value of J at which the rotational levels of the electronic state A begin to lie above those of the electronic state B.

Such irregularities in the position of the rotational levels are commonly called *perturbations* and have been found in numerous cases. According to the above remarks there must lie near the rotational level J at which an irregularity occurs another rotational level with equal J belonging to a different electronic state. This expectation has been confirmed by Dieke (133) in the spectrum of He₂ and by Rosenthal and Jenkins (218) in the spectrum of CN. The reverse, however, is not true. Two rotational levels with equal J lying close together do not always give rise to a perturbation. The other conditions which have to be fulfilled besides the equality of J have been discussed in a paper by Kronig (48), where the theory of perturbations was first given.

Closely related to the phenomenon of perturbations is that of *predissociation* discovered by Henri (164). If in Fig. 4 we represent the energy values of the various vibrational levels of two electronic states A and B by horizontal lines with the energy measured in the upward direction, they will converge, according to Art. 4, to limits PP and QQ which correspond to a dissociation of the molecule into separate atoms by an increase in the energy of nuclear vibration and which in general will be different for the two electronic states. Above each limit there lies a continuous region of energy values corresponding to dissociation of the molecule into separate atoms with excess kinetic energy.

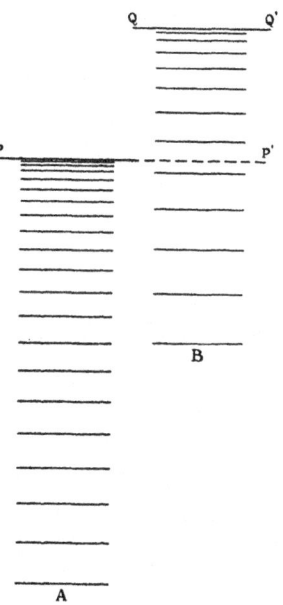

Fig. 4. Vibrational levels of two different electronic states A and B of a diatomic molecule. In the levels of B lying above PP' the molecule is in the state of predissociation.

From Fig. 4 it is apparent that some of the discrete levels of B lie in a region where the levels of A are already continuous, viz. those which have energies above PP. For these levels there exists then the possibility of the molecule dissociating spontaneously into separate atoms. The molecule in the discrete states of B lying above PP has thus a certain life-time, and if this life-time is comparable to the period of nuclear rotation, the rotational structure will become diffuse.

Such a diffuseness was first found by Henri and Teves (164) in the spectrum of S_2 [see also Henri and Wurmser (165)]. Bonhoeffer and Farkas (5) suggested the above explanation, and it was shown by Kronig (48) that the theoretical expectation for the *life-time of a molecule in the state of predissociation* under favourable conditions can actually be of the same order of magnitude as the period of nuclear rotation, while it may be expected to be large compared to the period of nuclear vibration.

12. Even and Odd Levels.

In Chap. I we concentrated our attention on the energy levels of a diatomic molecule, i.e. those values W for which the wave equation (2), Art. 1, or in the coordinates used, equation (11) or (19), Art. 2, has a solution Ψ, finite and single-valued in the domain of the independent variables. To be more precise, we investigated the values W^0 belonging to a differential equation for a function Ψ^0, (13) or (20), Art. 2, which differed by small terms from the actual wave equation. For the purposes of this chapter it will be necessary to discuss some properties of the *wave functions* Ψ themselves which have been investigated by Kronig (47), (48) and by Wigner and Witmer (85).

*If instead of the fixed coordinate system xyz from which we started in Art. 2 and with respect to which the position of the moving coordinate system $\xi\eta\zeta$ later introduced was referred we use another fixed coordinate system $x'y'z'$, oriented differently in space, then in the absence of external fields the wave equation expressed in the new coordinates will have exactly the same form as that written in the old. For there is nothing to distinguish the system $x'y'z'$ from the system xyz, all directions in space being equivalent. We can state this briefly by saying that *the wave* *equation is invariant against a change of orientation of the fixed coordinate system xyz.*

The function Ψ' obtained from a solution Ψ of the wave equation by the coordinate transformation which corresponds to going from the coordinate system xyz to the coordinate system $x'y'z'$ is itself a solution of the wave equation, belonging to the same energy value as Ψ. For, as we have seen, the transformed wave equation which it obeys is identical in form with the untransformed wave equation. By constructing the functions Ψ' corresponding to all possible orientations of the fixed coordinate system we obtain an infinite set of solutions of the wave equation, all belonging to the same energy value. These, however, are in general not linearly independent but will be linear combinations of a finite number of linearly independent solutions Ψ_1, Ψ_2, ... Ψ_g. g, the number of these linearly independent solutions, we shall call the *degree of spatial degeneracy* of the energy value under consideration.

Besides being invariant against a change of orientation of the fixed coordinate system xyz the wave equation will also be invariant against a substitution of left-handed coordinate systems for right-handed ones. For there is nothing in the nature of space to give preference to one or the other sort of coordinate system. In the derivations of Art. 2 we used two coordinate systems, a fixed one, xyz, and a moving one, $\xi\eta\zeta$. The latter was defined by its ζ-axis going through the two nuclei, let us say with the positive direction pointing from nucleus 2 to nucleus 1, while the ξ-axis was taken as lying in the xy-plane. The angle between the z- and the ζ-axis we called θ, where $0 \leqq \theta \leqq \pi$, and we chose the direction of the ξ-axis so that in turning from the z- to the ζ-axis through the angle θ we were going in the positive sense around the ξ-axis. The angle ψ from the positive x-axis to the projection of the positive ζ-axis on the xy-plane was also to be measured in the positive sense around the z-axis. Left-handed coordinate systems defined in an analogous "left-handed" fashion may be obtained by reversing the positive directions of all three fixed axes xyz, i.e. by "reflecting" them at the origin, by reversing the positive direction of the η-axis, and by measuring ψ in the negative direction around the new z-axis. The configuration of the molecule will then be specified by the new positional coordinates $\bar{\xi}_r, \bar{\eta}_r, \bar{\zeta}_r, \bar{\rho}, \bar{\theta}, \bar{\psi}$, related to the old by the transformation

$$\left. \begin{aligned} \bar{\xi}_r = \xi_r, \quad \bar{\eta}_r = -\eta_r, \quad \bar{\zeta}_r = \zeta_r, \\ \bar{\rho} = \rho, \quad \bar{\theta} = \pi - \theta, \quad \bar{\psi} = \psi + \pi, \end{aligned} \right\} \quad \dots\dots\dots\dots(1)$$

which for brevity we shall call a *reflection at the origin*.

If as in case (a) the spins are referred to the moving axes $\xi\eta\zeta$, then the spin operators will undergo the transformation

$$\bar{S}_{r\xi} = -S_{r\xi}, \quad \bar{S}_{r\eta} = S_{r\eta}, \quad \bar{S}_{r\zeta} = -S_{r\zeta}. \quad \dots\dots\dots(2)$$

For being the components of spin angular momentum they must transform in the same way as the components of orbital angular momentum $\mathcal{L}_{r\xi} = -i\,(\eta_r\,\partial/\partial\zeta_r - \zeta_r\,\partial/\partial\eta_r),\ \mathcal{L}_{r\eta},\ \mathcal{L}_{r\zeta}$. A component $\Psi(\bar{\sigma}_1, \dots \bar{\sigma}_r, \dots)$ of the wave function in the new coordinates is equal to that component $\Psi(\sigma_1, \dots \sigma_r, \dots)$ in the old coordinates in which in place of $\bar{\alpha}_r$ there stands β_r and in place of $\bar{\beta}_r$, α_r if the number of indices α is even, or equal to that component multiplied by -1 if the number of indices α is odd.

If as in case (b) the spins are referred to the fixed axes xyz, then the spin operators will not be changed by the transformation:

$$\bar{S}_{rx} = S_{rx}, \quad \bar{S}_{ry} = S_{ry}, \quad \bar{S}_{rz} = S_{rz}. \quad \dots\dots\dots\dots(3)$$

For they transform as $\mathcal{L}_{rx} = -i\,(y_r\,\partial/\partial z_r - z_r\,\partial/\partial y_r),\ \mathcal{L}_{ry},\ \mathcal{L}_{rz}$, the components of orbital angular momentum around the fixed axes, and these do not change when upon reflection $-x_r, -y_r, -z_r$ are substituted for x_r, y_r, z_r. A component $\Psi(\bar{s}_1, \dots \bar{s}_r, \dots)$ of the wave function is equal to the component $\Psi(s_1, \dots s_r, \dots)$ in which $\bar{s}_r = s_r$.

The equivalence of right- and left-handed coordinate systems as regards the form of the wave equation we may now state briefly by saying: *The wave equation is invariant against a reflection at the origin*, meaning thereby that on introducing new positional coordinates according to equations (1), new spin operators according to equations (2) or (3) and by transforming the components of the wave function as described above, the form of the wave equation remains unchanged.

The functions $\overline{\Psi}_1, \ldots \overline{\Psi}_g$ obtained by subjecting the solutions $\Psi_1, \ldots \Psi_g$ discussed previously to the process of reflection will thus also be solutions of the wave equation belonging to the same energy value as $\Psi_1, \ldots \Psi_g$. Now let us assume that $\Psi_1, \ldots \Psi_g$ are the only linearly independent solutions belonging to this energy value so that there is no other degeneracy besides that of spatial orientation. Then $\overline{\Psi}_1, \ldots \overline{\Psi}_g$ must be linear aggregates of $\Psi_1, \ldots \Psi_g$, or

$$\overline{\Psi}_i = \sum_{j=1}^{g} A_{ij} \Psi_j.$$

Regarding the A_{ij} as forming a matrix $\mathbf{A}$, we may write this as

$$\overline{\Psi} = \mathbf{A}\Psi.$$

According to a theorem of matrix algebra it is always possible to choose the Ψ's in such a way that $\mathbf{A}$ is a diagonal matrix, i.e. that only the elements A_{ii} are different from zero. For if that is not the case for the Ψ's originally given, all we have to do is to introduce a new set of Ψ's, say $\Psi_1', \ldots \Psi_g'$, which are linear combinations of $\Psi_1, \ldots \Psi_g$ so that

$$\Psi' = \mathbf{B}\Psi,$$

where $\mathbf{B}$ is another matrix. Then

$$\overline{\Psi}' = \mathbf{A}'\Psi' = \mathbf{B}\mathbf{A}\mathbf{B}^{-1}\Psi',$$

and there always exists a $\mathbf{B}$ so as to make $\mathbf{A}' = \mathbf{B}\mathbf{A}\mathbf{B}^{-1}$ diagonal.

Assuming then $\mathbf{A}$ as diagonal we conclude that its diagonal elements must be 1 or -1. For if we perform the process of reflection on the Ψ's twice in succession we get back the Ψ's we started from so that

$$\mathbf{A}^2 = 1,$$

where 1 is the unit matrix with diagonal elements equal to 1. A_{ii}^2 is thus equal to 1 and hence $A_{ii} = 1$ or -1 as asserted above. We shall finally prove that all the diagonal elements of $\mathbf{A}$ have the same sign. If we consider e.g. the function Ψ_1, then on changing the orientation of the coordinate system from xyz to $x'y'z'$ and calling the solutions resulting from $\Psi_1, \dots \Psi_g$ by the corresponding transformation $\Psi_1', \dots \Psi_g'$, we shall have

$$\Psi_1 = \sum_{j=1}^{g} C_{1j} \Psi_j' \quad\quad\quad\quad\quad\quad (4)$$

with constants C_{1j} depending on the change of orientation. Reflection at the origin gives

$$\overline{\Psi}_1 = \sum_{j=1}^{g} C_{1j} \overline{\Psi}_j'.$$

Now from $\overline{\Psi}_1 = A_{11} \Psi_1$ and from equation (4) we get

$$A_{11} \sum_{j=1}^{g} C_{1j} \Psi_j' = \sum_{j=1}^{g} C_{1j} \overline{\Psi}_j', \quad\quad\quad\quad (5)$$

where $A_{11} = \pm 1$. Since the set of constants C_{1j} may be continuously varied by changing the orientation of the coordinate system $x'y'z'$, and since equation (5) holds for all these orientations, we see that

$$\overline{\Psi}_j' = A_{jj} \Psi_j' = A_{11} \Psi_j'$$

so that all the A_{jj} are the same, equal either to 1 or -1.

The results obtained we may summarise as follows: By changing the orientation of the coordinate system xyz to which the motion of a diatomic molecule is referred, we may get in the absence of external fields from a solution Ψ of the wave equation other solutions belonging to the same energy value, all of which are linear aggregates of a finite number g of linearly independent solutions. If besides these there are no other linearly independent solutions belonging to the same energy value, either they do not change at all under the transformation called reflection or they all are multiplied by the factor -1. In the first case we call the energy level *even*, in the second case *odd*.

In the preceding chapter we classified the energy levels of a diatomic molecule by means of quantum numbers, and we now wish to investigate which of the levels there discussed are even and which are odd. For that purpose we shall first study the behaviour of the approximate wave functions Ψ^0, given by equations (13) and (20), Art. 2, under reflection, and we shall hence require more detailed information as to their exact form.

Let us first consider case (a), i.e. the wave function given by equation (13), Art. 2. It depends upon the electronic coordinates ξ_r, η_r, ζ_r, σ_r through the function Φ, obeying equation (14), Art. 2. This equation has the property of being invariant when the positional coordinates are transformed according to

$$\bar{\xi}_r = \xi_r, \quad \bar{\eta}_r = -\eta_r, \quad \bar{\zeta}_r = \zeta_r, \quad \bar{\rho} = \rho,$$

the spin operators according to

$$\bar{s}_{r\xi} = -s_{r\xi}, \quad \bar{s}_{r\eta} = s_{r\eta}, \quad \bar{s}_{r\zeta} = -s_{r\zeta},$$

and when $\Phi(\bar{\sigma}_1, \dots \bar{\sigma}_r, \dots)$ is taken equal to that component $\Phi(\sigma_1, \dots \sigma_r, \dots)$ in which in place of $\bar{\alpha}_r$ there stands β_r and in place of $\bar{\beta}_r$, α_r provided the number of indices α is even, or equal to that component multiplied by -1 provided the number of indices α is odd. For this transformation signifies that we are going from a right-handed coordinate system $\xi\eta\zeta$ to a left-handed one $\bar{\xi}\bar{\eta}\bar{\zeta}$ by taking the $\bar{\eta}$-axis in the reverse direction as the η-axis while leaving the other two axes unchanged, and for the description of the molecule with fixed nuclei the two systems are completely equivalent, the ζ-axis being an axis of symmetry. If Φ is a solution of equation (14), Art. 2, then the function $\bar{\Phi}$ resulting from it by the transformation just described will also be a solution belonging to the same energy value.

Now we have seen in Art. 3 that with any electronic level coming under case (a) for which the component of angular momentum along the internuclear line has the value $|\Omega|$ there coincides another level with a component of angular momentum $-|\Omega|$, which corresponds to reversing in the electronic motion the sense of rotation around the internuclear line. If we designate the wave function belonging to the first by $\Phi_{q|\Omega|}$, where q is an abbreviation for the quantum numbers required besides Ω to specify the electronic level, then the wave function $\Phi_{q-|\Omega|}$ of the second will be given by

$$\Phi_{q-|\Omega|} = \overline{\Phi}_{q|\Omega|},$$

i.e. it will be just the additional solution called for by the invariance of the equation for Φ. We shall normalise Φ, i.e. we shall multiply it by such a function of ρ that for all values of ρ

$$\sum_{\sigma_1, \ldots \sigma_r, \ldots} \int \Phi\Phi^* \, d\xi_1 \, d\eta_1 \, d\zeta_1 \ldots d\xi_r \, d\eta_r \, d\zeta_r \ldots = 1,$$

Φ^* denoting the conjugate of Φ, while in the summation each spin coordinate σ_r takes its two values α_r and β_r. The energy value $W_a'(\rho)$ of equation (14), Art. 2, belonging to the two functions $\Phi_{q|\Omega|}$ and $\Phi_{q-|\Omega|}$ we may designate by $W'_{q|\Omega|}(\rho)$.

The function Ψ^0 depends on the angles θ, ψ through the function Θ obeying equation (16), Art. 2. Now we have already mentioned in Art. 5 that the values of $W_a''(\rho)$ for which this equation has finite solutions are

$$W_a''(\rho) = \frac{h^2}{8\pi^2\mu\rho^2}[J(J+1)-\Omega^2], \quad J = |\Omega|, |\Omega|+1, \ldots (6)$$

and as shown by Reiche (80) and by Kronig and Rabi (46), the corresponding functions Θ are given by

$$\Theta_{J\Omega M} = \left[\frac{(-1)^{\alpha+\beta-\gamma}}{2\pi \, 2^{\alpha+\beta}} \frac{(\alpha-\beta)(\alpha-1)! \, (\gamma-\beta-1)!}{(-\beta)! \, [(\gamma-1)!]^2(\alpha-\gamma)!}\right]^{\frac{1}{2}}$$

$$\times (1+\cos\theta)^{\frac{\gamma-1}{2}} (1-\cos\theta)^{\frac{\alpha+\beta-\gamma}{2}} F[\alpha, \beta, \gamma, \tfrac{1}{2}(1+\cos\theta)]e^{iM\psi},$$

$$M = -J, -J+1, \ldots J, \ldots\ldots(7)$$

where

$$\alpha = \tfrac{1}{2}|M+\Omega| + \tfrac{1}{2}|M-\Omega| + J + 1,$$
$$\beta = \tfrac{1}{2}|M+\Omega| + \tfrac{1}{2}|M-\Omega| - J,$$
$$\gamma = |M+\Omega| + 1$$

and F is the hypergeometric function. They are normalised, i.e.

$$\int_0^\pi \int_0^{2\pi} \Theta\Theta^* \sin\theta \, d\theta d\psi = 1,$$

where Θ^* denotes the conjugate of Θ, and in addition

$$\Theta_{J\Omega M}(\pi - \theta, \psi + \pi) = (-1)^{J-\Omega} \Theta_{J-\Omega M}(\theta, \psi). \ldots\ldots(8)$$

We see from equation (7) that to a given J and Ω there belongs a set of $2J+1$ functions Θ.

The function $P(\rho)$ entering in Ψ^0 obeys equation (15), Art. 2. Since in this equation $W_a'(\rho)$ depends upon the quantum numbers q, $|\Omega|$, $W_a''(\rho)$ according to equation (6) upon the quantum numbers J, $|\Omega|$, P will also depend upon q, $|\Omega|$, J and besides upon a vibrational quantum number v, and so will W^0, the energy value. We may write hence $P_{q|\Omega|vJ}(\rho)$ and $W^0_{q|\Omega|vJ}$. Again we may normalise P in such a way that

$$\int_0^\infty PP^* \rho^2 d\rho = 1.$$

On the basis of the preceding considerations let us investigate the set of approximate wave functions belonging to one and the same energy level $W^0_{q|\Omega|vJ}$. There are $2(2J+1)$ linearly independent solutions, any linear combination of which will of course also be a solution. Let us take as solutions the $2(2J+1)$ linearly independent functions

$$\left.\begin{aligned}
\frac{1}{\sqrt{2}}(\Phi_{q|\Omega|}\, P_{q|\Omega|vJ}\, \Theta_{J|\Omega|M} + \overline{\Phi}_{q|\Omega|}\, P_{q|\Omega|vJ}\, \Theta_{J-|\Omega|M}),\\
\frac{1}{\sqrt{2}}(\Phi_{q|\Omega|}\, P_{q|\Omega|vJ}\, \Theta_{J|\Omega|M} - \overline{\Phi}_{q|\Omega|}\, P_{q|\Omega|vJ}\, \Theta_{J-|\Omega|M}).
\end{aligned}\right\} \quad ...(9)$$

According to equation (8) reflection will cause the upper set of the functions (9) to be multiplied by a factor $(-1)^{J-|\Omega|}$, the lower set by a factor $(-1)^{J-|\Omega|+1}$. For successive values of J the functions of the upper set are hence alternately even and odd, those of the lower set alternately odd and even.

In Art. 10, when taking into account the terms neglected in finding the approximate wave function Ψ^0, we have seen that each state with energy $W^0_{q|\Omega|vJ}$ actually is split into two states. The interaction of electronic motion, nuclear vibration, and nuclear rotation which these terms represent removes the degeneracy arising from the equivalence of the two senses of rotation around the internuclear line in the electronic motion of the molecule with fixed nuclei. Since then only the spatial degeneracy remains, the wave functions belonging to a component level must, according to the remarks made at the beginning of this article, be either all even or all odd. Moreover, when the perturbing terms are imagined reduced to zero adiabatically, the exact wave functions Ψ must go over into the approximate wave functions Ψ^0. Since during that process even wave functions remain even, odd ones odd, we can get information about the behaviour of the exact wave functions under reflection from the behaviour of the approximate wave functions.

We see thus that the wave functions of the one component level $q, |\Omega|, v, J$ must go over into the functions (9) with the $+$sign, those of the other component level into the functions (9) with the $-$ sign. In other words we have a pair of levels one of which is even, the other odd. For successive values of J the even and the odd level will lie alternately higher, as an exact analysis shows (48), and as may already be guessed at from the fact previously remarked that the functions of the upper set as well as of the lower set (9), are alternately even and odd.

Fig. 5 illustrates this result. On a horizontal line are indicated the successive rotational states with their quantum numbers J belonging to a given set of quantum numbers $q, |\Omega|, v$. The values J are indicated above. According to Art. 5 they are integral if the molecule has an even number of electrons, and half-integral

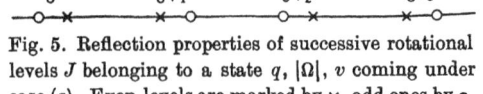

Fig. 5. Reflection properties of successive rotational levels J belonging to a state q, $|\Omega|$, v coming under case (a). Even levels are marked by ×, odd ones by o.

if the molecule has an odd number of electrons. Also the smallest value of J, as we have seen in Art. 5, is $|\Omega|$. Each rotational level is double, one component × being even, the other o odd. The distances are not supposed to represent energy differences, the diagram simply showing the reflection properties of the levels for different values of J. Since case (a) has been defined as corresponding to a large interaction of the spin, Σ-levels, for which this interaction is always minute, are to be classed as case (b) and singlet levels, for which there is no resultant spin at all, may also be properly considered as coming under that heading.

In case (b) we may proceed in a manner quite analogous. However, some special considerations are necessary here when $\Lambda = 0$, i.e. for Σ-states. For $\Lambda \neq 0$ we get instead of the functions (9) $2\,(2K+1)$ solutions

$$
\left.
\begin{aligned}
&\frac{1}{\sqrt{2}}(\Phi_{q|\Lambda|}\, P_{q|\Lambda|vK}\, \Theta_{K|\Lambda|M_K} + \bar{\Phi}_{q|\Lambda|}\, P_{q|\Lambda|vK}\, \Theta_{K-|\Lambda|M_K}), \\
&\frac{1}{\sqrt{2}}(\Phi_{q|\Lambda|}\, P_{q|\Lambda|vK}\, \Theta_{K|\Lambda|M_K} - \bar{\Phi}_{q|\Lambda|}\, P_{q|\Lambda|vK}\, \Theta_{K-|\Lambda|M_K}),
\end{aligned}
\right\}
\quad\ldots\ldots(10)
$$

belonging to the same energy value $W^0_{q|\Lambda|vK}$, which on reflection are multiplied by a factor $(-1)^{K-|\Lambda|}$ for the upper set, $(-1)^{K-|\Lambda|+1}$ for the lower set. Taking into account first those terms neglected in the wave equation which do not contain the spins, we get a splitting of each energy level $W^0_{q|\Lambda|vK}$ into two component levels (see Art. 10), one of which is even, the other odd, just as in case (a).

The functions Φ of case (b) are products of a function Φ' depending only on the positional coordinates ξ_r, η_r, ζ_r, ρ, and a function depending only on the spin coordinates s_r by reasoning quite analogous to that used in the mathematical section of Art. 3 for case (a) when the interaction terms of the spins were at first neglected. Again, to satisfy Pauli's exclusion principle, the spin function must be built up in such a way as to make the functions Φ antisymmetric against interchange of the coordinates of any two electrons. Due to the neglected interaction terms containing the spins an energy level originally single will be split into component levels in a manner similar to that discussed in the mathematical section of Art. 3 for case (a). The parts Φ' of the wave functions Φ characterising the component levels will be identical, and only the part depending on the spin coordinates will be different. This state of affairs may be conveniently described with the aid of a model by saying that the total spin S of the electrons is coupled with the total moment of momentum K of the molecule without spin to give a resultant

$$J = |K - S|, \ |K - S| + 1, \ \dots K + S.$$

Each of the two component levels spoken of above will then suffer a further splitting into $2S + 1$ levels if $K \geqq S$ or $2K + 1$ levels if $S \geqq K$. Since in case (b) the spin coordinates are not affected by reflection, all the levels resulting from the even component will be even, those resulting from the odd component odd.

For $\Lambda = 0$ we must remember that the degeneracy in the electronic motion around the internuclear line in the molecule with fixed nuclei is not present. The solution $\bar{\Phi}_{q0}$ for the electronic motion obtained by subjecting Φ_{q0} to reflection must hence not really be a new solution but equal to const. Φ_{q0}, and since by performing the substitution twice in succession we get back to the function we started from, the constant must be ± 1 so that

$$\bar{\Phi}_{q0} = \pm \, \Phi_{q0}. \qquad \dots\dots\dots\dots\dots(11)$$

Then to a given set of values q, $\Lambda = 0$, v, K there belong $(2K + 1)$ functions Ψ^0, viz.

$$\Phi_{q0} \, \mathrm{P}_{q0vK} \, \Theta_{K0M_K}, \quad M_K = -K, \dots K, \qquad \dots\dots\dots(12)$$

which according to equation (8) are on reflection, i.e. under the transformation (1), multiplied by a factor $(-1)^{\delta + K}$, where $\delta = 0$ or 1, depending on whether the upper or lower sign is valid in equation (11). For successive values of K the rotational levels of a Σ-state are thus alternately even and odd, the lowest level $K = 0$ being even for $\delta = 0$ and odd for $\delta = 1$. The question as to how many of the Σ-levels belonging to given levels of the separate atoms, resulting when ρ is made infinite, have $\delta = 0$ and how many $\delta = 1$ has been considered in detail by Wigner and Witmer (85) on the basis of group theory.

If we are not dealing with singlet Σ-states, i.e. if $S \neq 0$, then, as mentioned in Art. 10, S is again combined with K to give a resultant

$$J = |K - S|, \ |K - S| + 1, \ \dots K + S,$$

and a splitting into component levels will take place which are all even or all odd, depending on whether the original level was even or odd.

Fig. 6 illustrates our results for case (b) for the various values of Λ and S, even levels being designated by ×, odd levels by o.*

The non-mathematical reader will have to take for granted that in diatomic molecules, depending on the nature of their wave functions, there are two kinds of levels, even and odd, and can inform himself about their occurrence from Fig. 5 for the rotational states of electronic levels coming under case (a) and from Fig. 6 for the rotational states of electronic levels coming under case (b).

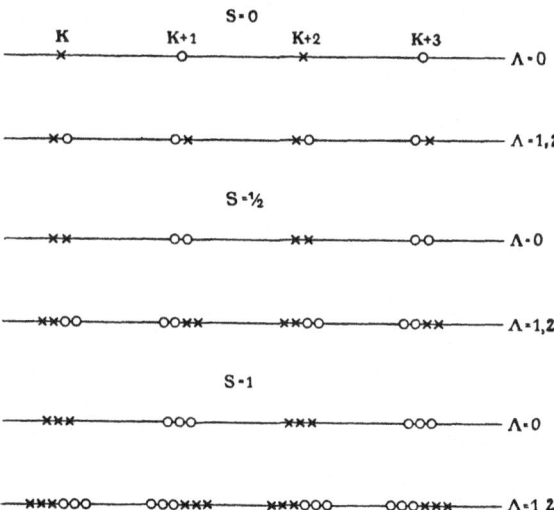

Fig. 6. Reflection properties of successive rotational levels K belonging to a state q, $|\Lambda|$, v coming under case (b). Even levels are marked by ×, odd ones by o.

13. SYMMETRICAL AND ANTISYMMETRICAL LEVELS.

If a diatomic molecule is *homonuclear*, as e.g. H_2, there exists besides the classification of the energy levels into even and odd ones another classification which too is based on the properties of the wave functions, and which was first introduced by Heisenberg [23] and Hund [31].

For molecules of this kind the wave equation, besides being invariant under the transformations described in the last article, *will also be invariant under the transformation signifying an interchange of the two nuclei,* the nuclei being in no way distinguishable. For the positional coordinates used this transformation has the form

$$\tilde{\xi}_r = - \xi_r, \quad \tilde{\eta}_r = \eta_r, \quad \tilde{\zeta}_r = - \zeta_r \atop \tilde{\rho} = \rho, \quad \tilde{\theta} = \pi - \theta, \quad \tilde{\psi} = \psi + \pi \Bigg\} \quad \ldots\ldots\ldots\ldots(1)$$

On the spin operators and the components of Ψ the transformation (1), both in case (a) and in case (b), has the same effect as the transformation (1) discussed in the previous article, as the reasoning there employed shows.

By quite analogous arguments we conclude that if there are no other linearly independent solutions of the wave equation associated with a given energy value besides the g wave functions corresponding to the spatial degeneracy of the stationary state, the latter do not change at all under the transformation signifying an interchange of the two nuclei or are all multiplied by the factor -1. In the first case we call the energy level *symmetrical in the nuclei,* in the second case *antisymmetrical in the nuclei* or for brevity, when no misapprehension is possible, simply *symmetrical* and *antisymmetrical.*

The procedure for investigating the properties of the wave function Ψ with respect to the interchange of the nuclei is exactly analogous to that used in the preceding article when studying their behaviour under reflection. In fact the expressions (9), (10) and (12) for the wave functions and the equations (8) and (11), Art. 12, together with the whole argumentation based on them, may be taken over literally if only we replace Φ by $\tilde{\Phi}$ and substitute the terms "interchange of nuclei, symmetrical and antisymmetrical" for the terms "reflection, even and odd." Here $\tilde{\Phi}$ denotes the function resulting from Φ when the positional coordinates are transformed according to

$$\tilde{\xi}_r = - \xi_r, \quad \tilde{\eta}_r = \eta_r, \quad \tilde{\zeta}_r = - \zeta_r, \quad \tilde{\rho} = \rho,$$

and when $\tilde{\Phi}(\tilde{\sigma}_1, \ldots \tilde{\sigma}_r, \ldots)$ is taken equal to that component $\Phi(\sigma_1, \ldots \sigma_r, \ldots)$ in which in place of $\tilde{\alpha}_r$ there stands β_r and in place of $\tilde{\beta}_r$, α_r provided the number of indices α_r is even, or equal to that component multiplied by -1 provided the number of indices α_r is odd.

From this it follows specially that among the multiplet, vibrational, rotational, and fine structure levels of one and the same electronic state belonging to a homonuclear molecule all the even ones have the same kind of symmetry, all the odd ones the opposite kind. If the even levels are symmetrical, the odd ones are anti-symmetrical and vice versa. For either

$$\tilde{\Phi} = \bar{\Phi}, \quad \ldots\ldots\ldots\ldots\ldots\ldots\ldots(2)$$

or

$$\tilde{\Phi} = -\bar{\Phi}. \quad \ldots\ldots\ldots\ldots\ldots\ldots\ldots(3)$$

Since the functions P and Θ behave exactly in the same way whether we subject the wave functions (9), (10), and (12), Art. 12, to the transformation of reflection or the corresponding expressions with $\tilde{\Phi}$ instead of $\bar{\Phi}$ to the transformation of interchanging the nuclei, even levels will be symmetrical if equation (2) applies and antisymmetrical if equation (3) applies to the electronic state in question. Which of the two equations is valid for the various molecular electronic states arising from given states of the separate atoms has been discussed by Wigner and Witmer [85] with the help of group theory.*

Again the non-mathematical reader must take it for granted that in homonuclear diatomic molecules depending on the nature of their wave functions the energy levels fall into two classes, symmetrical and antisymmetrical, and is referred for the rule governing their occurrence to what has been said in the last paragraph of the preceding mathematical discussion.

CHAPTER III

SELECTION RULES AND INTENSITIES IN DIATOMIC MOLECULES

14. GENERAL FOUNDATIONS.

Up to the present we have only been concerned with the energy levels of a diatomic molecule and the functions Ψ, entering in equation (2), Art. 1, which according to wave mechanics are associated with them. As we have seen in Art. 1, the frequency of a line present in the spectrum of a diatomic molecule is connected with two of its energy values W_j and $W_{j'}$ by equation (1), Art. 1. We speak of the line as being due to the *transition* of the molecule between the states j and j'. Equation (1), Art. 1, tells us which frequencies *may* occur in the spectrum but not which *do* occur, nor what the *intensities* of the lines actually present will be.

According to the notions of the quantum theory the intensity of a line emitted spontaneously by an assemblage of atomic systems under suitable conditions of excitation depends upon two factors: (1) the number of systems N_j present in the higher one, W_j, of the two energy levels W_j and $W_{j'}$ to which, according to equation (1), Art. 1, the frequency $\nu_{jj'}$ of the line under consideration is related; (2) the *probability of spontaneous emission* $A^j_{j'}$ belonging to the pair of states j and j', as first introduced into the theory by Einstein †. The total amount of radiation of frequency $\nu_{jj'}$ emitted in all directions per unit time by the atomic systems will in fact be given by

$$N_j\,A^j_{j'}\,h\nu_{jj'}. \qquad\qquad\qquad (1)$$

Similarly the amount of energy of frequency $\nu_{jj'}$ absorbed by the assemblage when outside radiation is falling on it depends (1) on the number of systems $N_{j'}$ present in the lower energy level $W_{j'}$, (2) on the *probability of absorption* $B^j_{j'}$ belonging to the two states j and j', also first defined by Einstein. If the incident radiation is

† Einstein, *Phys. Zeit.* **18**, 121, 1917.

coming equally from all directions and has a continuous range of frequencies with an energy density u in the neighbourhood of the frequency $\nu_{jj'}$, the energy of this frequency absorbed per unit time by the atomic systems will be

$$N_{j'} \, B^j_{j'} \, h\nu_{jj'} \, u. \quad\text{......................}(2)$$

The quantities N_j and $N_{j'}$ depend upon the state of the emitting or absorbing gas, i.e. on such factors as its temperature or the voltage and current density of an electric discharge to which it may be subjected. The quantities $A^j_{j'}$ and $B^j_{j'}$ on the other hand are characteristic of the atomic systems of which the gas is composed. By considering the temperature equilibrium between the atomic systems and black-body radiation, Einstein showed quite generally that

$$B^j_{j'} = \frac{c^3}{8\pi \, h\nu^3_{jj'}} \frac{g_j}{g_{j'}} A^j_{j'}, \quad\text{......................}(3)$$

where g_j and $g_{j'}$ are the statistical weights of the states j and j', so that if $A^j_{j'}$ is known $B^j_{j'}$ may be computed. It is the object of this chapter to determine $A^j_{j'}$ for diatomic molecules.

The wave mechanics permits the calculation of $A^j_{j'}$ if the wave functions Ψ_j and $\Psi_{j'}$ of the two states j and j' are known. For it expresses $A^j_{j'}$ in first approximation in terms of the matrix element $\mathfrak{P}(j; j')$ of the dipole moment of the molecule by means of the formula

$$A^j_{j'} = \frac{64\pi^4 \nu^3_{jj'}}{3c^3 h} \mathfrak{P}(j; j') \mathfrak{P}(j'; j), \quad\text{............}(4)$$

and this matrix element is given by

$$\mathfrak{P}(j; j') = \sum_s \int \Psi_j^* \sum_r e_r \, \mathfrak{r}_r \, \Psi_{j'} \, d\tau, \quad\text{...............}(5)$$

e_r denoting the charge, $\mathfrak{r}_r$ the radius vector of the rth particle in the atomic system, while the summation $\sum\limits_r$ extends over all the particles, the integration over the domain of the positional co-ordinates, and the summation $\sum\limits_s$ over all values of the spin coordinates. In further approximation quadrupoles and higher poles have to be taken into account.

It may happen that $\mathfrak{P}(j; j')$ vanishes. Then, according to equations (4) and (3), the corresponding line does not occur in the spectrum. Thus, as we shall see in the subsequent articles, $\mathfrak{P}(j; j')$ is zero for any pair of states of a diatomic molecule whose total angular momentum J differs by more than unity, provided external fields are absent. If, as in the example just mentioned, a large class of transitions is excluded, we speak of a *selection rule*. We shall see presently that for diatomic molecules the number of transitions corresponding to all possible pairs of energy values W_j and $W_{j'}$ is greatly reduced by selection rules. This must be regarded as a very fortunate circumstance, for otherwise the spectrum would show such an abundance of lines that an analysis might become impossible or at least be greatly hampered.

15. ELECTRONIC BANDS.

We consider, to start with, those transitions which lead from one electronic level of a diatomic molecule to a different one. The associated frequencies usually lie in the visible or ultra-violet part of the spectrum. We shall at first simply state what is known regarding the occurrence of such transitions and their intensities, postponing the mathematical derivation of the results.

1. *In all diatomic molecules even levels combine only with odd ones and vice versa; transitions between two even levels or two odd levels do not occur.*

This rule, which was obtained by Kronig (47), (48) and by Wigner and Witmer (85), has been especially tested by Bengtsson and Hulthén (195) for the bands of AlH. It is derived under the assumption that the radiation emitted by the molecule may be calculated from its dipole moment, the radiation due to quadrupoles and higher poles being negligible in comparison, and also that external fields are absent. The first supposition is quite justified since the ratio of the quadrupole moments to the dipole moments is of the same order of magnitude as the ratio of the atomic dimensions to the wave-length of the radiation emitted, or at most about 10^{-3} for the optical region of the spectrum. The ratio of the quadrupole radiation to the dipole radiation, which manifests itself in the ordinary kind of transitions, is proportional to the square of this number, i.e. 10^{-6}, so that transitions due to quadrupole radiation would in general be much too rare to be observable in the spectrum. Any deviations from the rule, such as those observed by Watson [187] in the bands of OH, must hence be regarded as due to external fields.

2. *In homonuclear diatomic molecules symmetrical levels combine only with symmetrical ones, antisymmetrical levels only with antisymmetrical ones, while transitions between symmetrical and antisymmetrical levels are forbidden.*

This rule, which is due to Heisenberg [23] and Hund [31], would be valid even if the radiation due to quadrupoles and higher poles were taken into account and if external fields were present so that it should also apply to transitions caused by impacts. This general validity was indeed confirmed through a striking experiment performed by Wood and Loomis [176]. By illuminating iodine vapour with a mercury arc they were able to excite one particular rotational level ($J = 34$) of a certain vibrational level belonging to an excited $^1\Sigma$-state of the molecule I_2. In the resonance radiation there appeared lines corresponding to transitions starting from this rotational level. Besides these, however, the resonance radiation also contained frequencies corresponding to transitions starting from the neighbouring rotational levels with even J, but not with odd J. Since for a $^1\Sigma$-state, according to Art. 13, all levels with even J have the same kind of symmetry, those with odd J the opposite kind, we can interpret the experimental result as being due to the transfer by impact of I_2-molecules from the level $J = 34$ originally excited to the other rotational levels with even J before re-emission had taken place, while transfers to the rotational levels with odd J were forbidden by our selection rule.

In the derivation of the rule the nuclei are regarded as point charges, and the terminology "symmetrical, antisymmetrical" means that the wave functions are symmetrical and antisymmetrical in the positional coordinates of the two equal nuclei. In Art. 18 we shall see that some nuclei have a spin, which has the effect that the wave functions are no longer rigorously symmetrical or anti-symmetrical in the positional coordinates of the nuclei but only approximately so. However, due to the smallness of the interaction energy by which the nuclear spin is coupled to the molecule, the deviations of the wave functions from strictly symmetrical or anti-symmetrical functions are very minute, and in consequence transitions at variance with our selection rule will be so rare as not to be observable in the spectrum. That they actually may take place we shall see in Art. 25.

From a joint application of rules 1 and 2 and with the aid of the results of Arts. 12 and 13 it may be concluded that if in a homonuclear molecule there are three electronic states 1, 2, 3, and if there are radiative transitions between the rotational levels of 1 and 2 and of 1 and 3, then no radiative transitions can take place between the rotational levels of 2 and 3, a result which seems to be confirmed by experience.

3. *Electronic levels of different multiplicity do not combine.*

In the derivation of this rule which, together with rules 4, 5, 6 is due to Hund (29), (31), it is presupposed that the wave functions belong to a definite class of symmetry as regards the interchange of the positional coordinates of the electrons. Similarly as under 2, this is true only as long as the interaction energy by which the electron spins are coupled to the molecule is vanishingly small. When this is not the case, the transitions forbidden by rule 3 will appear with an intensity whose ratio to that of the lines not for-bidden will be of the order of magnitude $\Delta\nu/\nu$, $\Delta\nu$ being the frequency difference of two component levels of that one of the two electronic states which has the larger multiplet intervals, ν the frequency of the line emitted. Since in general $\Delta\nu/\nu$ is of the order 10^{-3} or less, transitions between electronic levels of different multiplicity will indeed be rare. An example of a violation of our rule is the atmospheric absorption bands of O_2, which represent a transition from the normal $^3\Sigma$-state to a higher $^1\Sigma$-state of that molecule. Due to the thickness of the absorbing layer, the atmosphere, these bands are noticeable in spite of the small transition probability.

4. *Only such electronic levels combine for which* $|\Lambda|$ *does not differ by more than unity.*

This means that Σ-states combine only with Σ- and Π-states, Π-states only with Σ-, Π- and Δ-states, etc. On account of the interaction energy by which the spins are coupled to the molecule and on account of the molecular rotation exceptions to this rule will occasionally but rarely be found.

5. *For transitions between two electronic levels which both may be considered as coming under case* (a) Σ *remains unchanged.*

We can expect this rule to hold only approximately in practice since the condition of case (a), viz. that the effect of the interaction terms by which the spins are coupled to the molecule with fixed nuclei are large compared to the influence of nuclear rotation, is never very well satisfied.

6. *For transitions between two electronic levels which both may be considered as coming under case* (b) K *does not change by more than unity.*

For the reason that the condition of case (b), viz. that the effect of the interaction terms by which the spins are coupled to the molecule with fixed nuclei are small compared to the influence of nuclear rotation, is sometimes not well satisfied this rule too has a limited validity only.

7. *The total angular momentum* J *cannot change by more than unity.*

Taking only the dipole radiation into account and assuming that external fields are absent this rule should hold rigorously.

Proofs. To 1. Under the transformation called reflection the quantity $\Sigma_r e_r \mathfrak{r}_r$ entering in equation (5), Art. 14, reverses sign. If now the two wave functions Ψ_j and $\Psi_{j'}$ both remain unchanged or both reverse sign under this same transformation, i.e. according to Art. 12 if the levels j and j' are both even or both odd, then the right-hand side of equation (5), Art. 14, will become equal to its negative under the transformation so that $\mathfrak{P}(j; j')$ and hence according to equation (4), Art. 14, $A^j_{j'}$ is zero. In the presence of external fields the functions Ψ do not retain the property of remaining unchanged except for their sign by the process of reflection, since the invariance of the wave equation under this transformation will thereby in general be destroyed. Therefore the selection rule may be invalidated by the action of external fields.

To 2. Interchange of the two equal nuclei leaves the quantity $\Sigma_r e_r \mathfrak{r}_r$ in equation (5), Art. 14, and also the corresponding quantities in the matrix elements of quadrupoles and higher poles, unaltered. If one of the two wave functions Ψ_j and $\Psi_{j'}$ remains unchanged during the corresponding transformation while the other reverses sign, i.e. according to Art. 13 if one of the levels is symmetrical, the other antisymmetrical in the positional coordinates of the nuclei, then again $\dot{\mathfrak{P}}(j; j')$ will vanish for the same reason as under 1, and thus also $A^j_{j'}$ according to equation (4), Art. 14. An external field will not destroy the invariance of the wave equation against interchange of two equal nuclei since it affects them both in the same way so that the wave functions Ψ retain their symmetry properties and the above proof remains valid even in the presence of such a field.

The rule at the end of 2 may easily be seen to hold by the following arguments. According to Art. 13, if one of the even rotational levels belonging to a given electronic state in a homonuclear molecule has a certain kind of symmetry, say is antisymmetrical, all other even rotational levels belonging to the same electronic state have the same kind of symmetry, while the odd levels have the opposite kind of symmetry, being symmetrical in our example. If this electronic state, which we may call 1, is to combine with other electronic states 2 and 3, then according to selection rules 1 and 2 even rotational levels both for states 2 and 3 must be symmetrical, odd ones antisymmetrical. But then according to the same selection rules no combinations can take place between the levels of states 2 and 3. Similar arguments apply if the even levels of state 1 are symmetrical in place of anti-symmetrical.

To 3. Not only the interchange of the positional coordinates of two equal nuclei but also the interchange of the positional co-ordinates of two electrons leaves the quantity $\sum_r e_r \mathfrak{r}_r$ in equation (5), Art. 14, unaltered. Now we have seen in Art. 3 that as long as the interaction of the electron spins with the rest of the molecule is neglected, each wave function is simply a product of a function depending on the positional coordinates alone and another function depending solely on the spin coordinates. As discussed in detail by Wigner and Witmer (85), the function of the positional coordinates belongs to a certain class of symmetry as regards the interchange of the electrons, and the multiplicity of the corresponding state is thereby determined. By investigating the behaviour of the right-hand side in equation (5), Art. 14, when the electrons are inter-changed, it can be shown by arguments which are a generalisation of those used in the preceding proof that the integral may be made equal to its negative and will hence vanish whenever the parts of Ψ_j and $\Psi_{j'}$, depending only on the positional coordinates, do not belong to the same class of symmetry, i.e. when the states j and j' do not have the same multiplicity. Due to the fact, however, that the splitting of Ψ into two parts depending respectively on the positional and the spin coordinates is only possible approximately on account of the interaction energy by which the spins are coupled to the molecule, transitions at variance with selection rule 3 may occur with small probability.

To 4. In Art. 3 we have seen that in the molecule with fixed nuclei and with the influence of the spins neglected, positional coordinates could be introduced such that one of them, ϕ_1, occurred in the wave function belonging to a state of this system only

through the factor $e^{\pm i\Lambda\phi_1}$. Now the terms of $\Sigma_r e_r \mathfrak{r}_r$ in equation (5),
Art. 14, will contain ϕ_1 in the combinations e^{ϕ_1}, $e^{-\phi_1}$ or not at all.
The integration over ϕ_1 from 0 to 2π in that equation therefore
gives only then something different from zero if the values of Λ for
the two states j and j' do not differ by more than unity. This result
is not quite accurate because the electron spins and the rotation of
the molecule have the effect of destroying the exact dependence of
Ψ on ϕ_1 through the factor $e^{\pm i\Lambda\phi_1}$.

To 5. This selection rule too is based on the approximate wave
functions of the molecule with fixed nuclei which are products of
two functions depending on the positional and the spin coordinates
respectively. The result follows from the fact that $\Sigma_r e_r \mathfrak{r}_r$ in equation (5), Art. 14, does not contain the spins at all, but only positional
coordinates. Through the influence of rotation of the molecule the
coupling of the spins to the internuclear line will be loosened and
when the two influences become of equal importance the selection
rule will fail altogether.

To 6 and 7. While rule 7 is a general quantum mechanical result
according to which for any atomic system in isotropic space only those
matrix elements of the dipole moment $\mathfrak{P}$ $(j; j')$ do not vanish which
belong to two states j and j' with a total angular momentum J
differing by not more than unity, rule 6 really is this same result
for the orbital angular momentum K of a molecule in which the
interaction energy of the spins has been reduced to zero. Due to
the slight coupling energy still actually present in case (b) the
validity of rule 6 is approximate only.*

Having formulated the various selection rules limiting the
possibilities of transition in diatomic molecules we come now to
discuss the *probability of* those *transitions* which are permitted by
the selection rules. In this discussion we shall in general confine
ourselves to using in the expression (5), Art. 14, which according
to equation (4), Art. 14, determines the probability of transition,
the approximate wave functions Ψ^0 of Chap. I, or in case the

Λ-doubling is not neglected, the linear combinations built up from them in Art. 12, instead of the exact wave functions Ψ. Again we state at first the results.

8. To make theoretical predictions as to the absolute or relative probability of transitions between different pairs of electronic levels seems hardly possible. For this would require a detailed knowledge of that part Φ of the wave functions Ψ^0 which depends upon the electronic coordinates, and such knowledge is usually lacking. We therefore limit ourselves to considering the relative probabilities of the transitions between the various multiplet, vibrational, and fine structure levels of one and the same pair of electronic states, which according to selection rule 4 have quantum numbers Λ and Λ' differing by not more than unity and according to selection rule 3 the same multiplicity $2S + 1$.

9. Imagining at first the multiplet structure, the rotational structure, and the fine structure (Λ- and ρ-doubling) reduced to zero so that only the vibrational structure of the two electronic levels remains, we investigate the relative probabilities of the doubly infinite set of transitions possible between their vibrational levels. Following a suggestion of Franck [19], Condon [9], [10], [11] has shown that if the intensities of these transitions are entered in a square array with the vibrational quantum number of the initial state v, say, put down in the horizontal direction, that of the final state v' in the vertical direction, then the most intense transitions lie in the neighbourhood of a parabola with an axis roughly in the direction of the diagonal of the square. The form of the parabola depends upon the shape of the potential energy curves governing the nuclear vibration in the two electronic states. Fig. 7 shows the transitions between the vibrational levels of the two electronic states of the molecule Na_2 discussed in Art. 3 as observed by Loomis and Wood [162], the numbers representing estimated intensities. The grouping of the more intense transitions around a curve of parabolic type is quite evident.

10. We now consider two particular vibrational levels belonging

to two different electronic states, both coming under case (a). If we imagine the multiplet structure introduced, each of the two vibrational levels will be resolved into component levels, the number of these component levels being the same for both, since according to rule 3 their multiplicity must be identical in order that combinations may take place at all. According to rule 5 every one of the component levels of the higher vibrational level combines only with

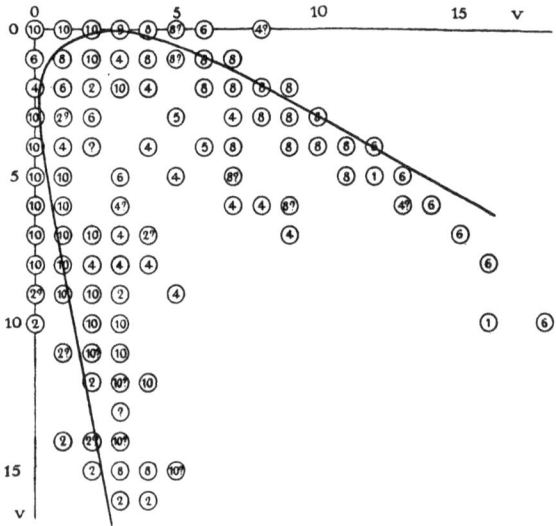

Fig. 7. Calculated locus of most probable vibrational transitions in the blue-green band system of Na_2. Numbers in circles give the observed intensities.

one component of the lower vibrational level, viz. that one which has the same Σ. The transition probability is the same for the transitions joining the various pairs of component levels with equal Σ.

11. Let us next take such a pair of combining component levels with equal Σ but with values Ω and Ω' of $\Omega = \Lambda + \Sigma$, which according to rule 4 may be equal or differ by unity. Due to the nuclear rotation we get from each an infinite set of rotational levels distinguished by the quantum number J, the smallest value of which for the two states is $J = |\Omega|$ and $J = |\Omega'|$ respectively.

According to rule 7 J must remain unaltered or change only by unity, and it is shown in a number of investigations [Fowler (18), Dieke (16), Hönl and London (27), Kemble (35), Mensing (61), Fues (21), Oppenheimer (78), Dennison (14), Kronig and Rabi (46), Rademacher and Reiche (79)] that the products $\mathfrak{P}(j; j')$ $\mathfrak{P}(j'; j)$ entering in equation (4), Art. 14, are given for the allowed transitions by the following expressions:

for $\Omega' = \Omega \pm 1$, $J' = J + 1$ by

$$\text{const. } O(J+1, \ \pm \Omega + 1)/2(J+1)(2J+1);$$

$$J' = J,$$

$$\text{const. } P(J, \ \pm \Omega)/2J(J+1);$$

$$J' = J - 1,$$

$$\text{const. } O(J+1, \ \mp \Omega - 1)/2J(2J+1);$$

$$\Omega' = \Omega, \ J' = J + 1,$$

$$\text{const. } Q(J+1, \ \Omega)/(J+1)(2J+1);$$

$$J' = J,$$

$$\text{const. } Q(\Omega, 0)/J(J+1);$$

$$J' = J - 1,$$

$$\text{const. } Q(J, \ \Omega)/J(2J+1);$$

where the primed state is the one of smaller energy and where

$$O(x, y) = (x + y)(x + y - 1), \quad \dots\dots\dots\dots(1)$$
$$P(x, y) = (x + y + 1)(x - y), \quad \dots\dots\dots\dots(2)$$
$$Q(x, y) = x^2 - y^2. \quad \dots\dots\dots\dots\dots\dots(3)$$

In each set of the three expressions for a given change of Ω the constant is common to all transitions belonging to the set, in other words it is independent of J. In all three sets addition of the three expressions gives simply this constant. The physical significance of this result is that the total transition probability from a rotational level J of the state Ω to all the rotational levels J' of the state Ω' is not influenced by the rotation and hence independent of J

(summation rule). Moreover, it follows from the above that one and the same constant must be used for the transitions between the various pairs of component levels of two multiple electronic states discussed under 10 in order that the result there stated may hold.

12. On account of the phenomenon of Λ-doubling discussed in Art. 10 every rotational level J of the last section consists of two fine structure components, of which according to Art. 12 one is even, the other odd. Instead of one transition from the rotational level J to the rotational level J' we shall in reality have two transitions if we take into consideration selection rule 1, viz. the transition from the even component of J to the odd component of J' and from the odd component of J to the even component of J'. The transition probabilities for these two transitions are equal.

13. We next consider two particular vibrational levels of two different electronic states both coming under case (b). In order that transitions may take place at all, the quantum numbers Λ and Λ' of the two states must, according to rule 4, not differ by more than unity. On each of the vibrational levels there is built up an infinite sequence of rotational levels distinguished by the quantum number K, the smallest value of which for the two states is $|\Lambda|$ and $|\Lambda'|$ respectively. Again, according to rule 6, K may not change by more than unity, and the quantities $\mathfrak{P}(j; j')\,\mathfrak{P}(j'; j)$ determining the relative probabilities of the allowed transitions will be given by expressions of exactly the same form as those in section 11 if only we replace there J by K and Ω by Λ. The resulting formulae have been tested by Ornstein and Van Wijk (149) for the negative nitrogen bands, which are due to a transition $^2\Sigma$—$^2\Sigma$ of the $N_2{}^+$-molecule.

14. We have seen in Art. 10 that if $S \neq 0$ the states K discussed in the last section show a fine structure, which may be described by saying that K and S are coupled to give a resultant J taking the values $|K - S|, |K - S| + 1, \ldots K + S$ (ρ-"doubling"). In consequence each of the transitions is in reality multiple. According to rule 7 J, just like K, may not change by more than unity, and the quantities $\mathfrak{P}(j; j')\,\mathfrak{P}(j'; j)$ determining the relative probabilities

of the different possible component transitions, as shown by Mulliken (71) and by Hill and Van Vleck (26), have exactly the same form as for multiplets in atomic spectra and are obtained from the expressions valid there by replacing the orbital angular momentum of the electrons L by K.

15. If a state J or J' of the last section belongs to an electronic state with $\Lambda \neq 0$, it will show Λ-doubling for high rotational quantum numbers, one of the component levels being even, the other odd. Again for the transitions between J and J' rule 1 applies. If the rule permits just one transition between J and J' (which will be the case unless both J and J' have Λ-doubling), the transition probability is to be obtained as described in the last section. If there are two component transitions, the transition probability for each component will be half that determined by the rule of the last section for the total probability of a transition from J to J'.

Proofs. To 9. The distribution of intensities among the various vibrational transitions belonging to two electronic states can be most conveniently understood with the help of the model used in the older quantum theory, which is here quite adequate since the result is purely qualitative.

Fig. 8 shows the potential energy functions governing the nuclear vibration, which we have discussed already in Art. 3 for two electronic states in a hypothetical case. The energy values corresponding to the various vibrational levels are indicated on the vertical scales laid off from the minimum point of each curve. Suppose that initially the molecule has one quantum of vibrational energy. In the model that means that the vibration takes place between those values of the internuclear distance ρ at which the curve marked 'initial" is intersected by the dotted line $W = $ const. through the energy value of its first vibrational level. Most of the time the vibrator spends at these extreme positions ρ_{min} or ρ_{max}, and if the molecule then undergoes an electronic transition by which the function governing the nuclear vibration is changed to the curve marked "final," the nuclei will most likely continue their motion in such a way that one of the new extreme positions ρ'_{min} and ρ'_{max} coincides with one of the original ones, ρ_{min} or ρ_{max}. This is indicated in the figure by drawing vertical lines through $\rho = \rho_{min}$ and $\rho = \rho_{max}$ and finding their intercepts with the final curve. The horizontal lines through these new intercepts give then the energy values in whose neighbourhood the final vibrational energy will most probably fall. There results in general from this construction the parabolic distribution mentioned in section 9 and confirmed by the experimental evidence.

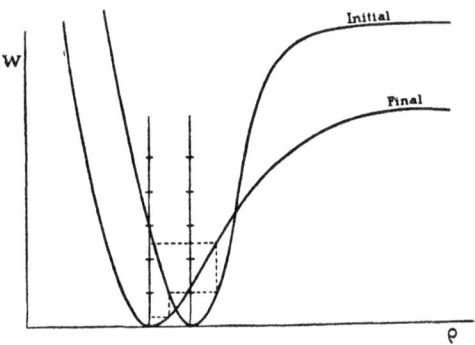

Fig. 8. Potential energy curves of diatomic molecule illustrating graphical method of finding favoured transitions.

It will be seen from Fig. 8 that for large vibrational quantum numbers of the initial state it may happen that $\rho'_{min} = \rho_{min}$ corresponds to a vibrational energy of the final state larger than the dissociation energy of the molecule. This means that the electronic transition will with great probability be accompanied by a tearing apart of the nuclei with the excess energy as kinetic energy. This state of affairs seems actually to occur for the molecules Cl_2, Br_2, and ICl as discussed in detail by Condon (11).

To 10, 11, 12, 13, 15. The results of these sections follow from substituting in the expression (5), Art. 14, for $\mathfrak{P}(j; j')$ the wave function Ψ^0 as described in Chap. II. The reader is referred for the details of the calculations to the articles quoted in the text.

To 14. That the formulae valid for atomic multiplets apply to the multiplets considered in this section if only the orbital angular momentum of the electrons in the atom L is replaced by the quantum number K, is readily seen. For in the atomic case the radiating system has an angular momentum L slightly coupled to the spin S and precessing with it about their resultant J, while in the case of the molecule K plays exactly the rôle of L.*

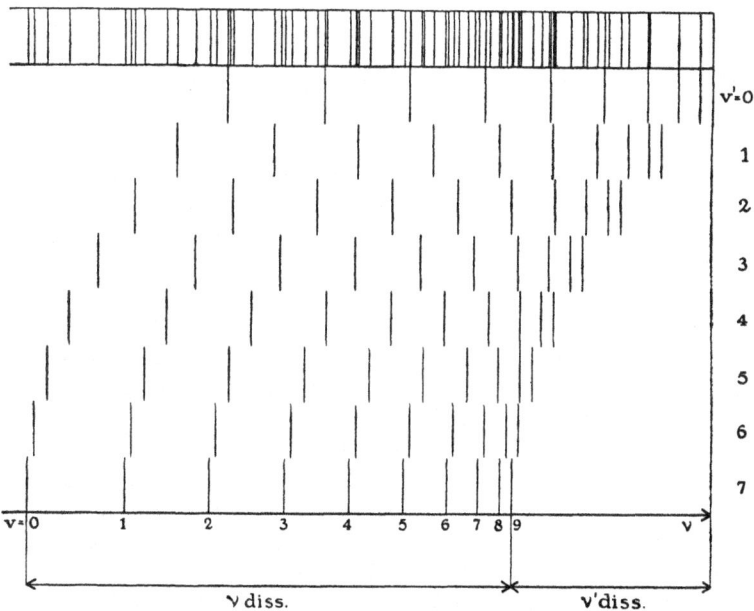

Fig. 9. Generation of a band system from the transitions between the vibrational levels v and v' of two different electronic states.

We are now in a position to discuss the actual *appearance of electronic bands* as they are obtained on spectrograms. All band lines belonging to the same electronic transition are said to form a *band system*. If for the present we disregard the multiplet, rotational, and fine structure of the electronic levels and retain only the vibrational structure, we can represent the totality of the vibrational transitions by the scheme of Fig. 9. Here the abscissa measures the frequency, and in each horizontal row all the vibrational transitions which end on the same lower vibrational level $v^* = 0, 1, 2, \ldots$ are indicated as vertical lines corresponding from left to right to values $0, 1, 2, \ldots$ of the vibrational quantum number v of the upper state. The distance between two lines v_1 and v_2 in the same row represents hence the frequency difference of the vibrational levels v_1 and v_2 of the upper state. Since for large values of v these levels converge towards a limit, as we saw in Art. 4, corresponding to the dissociation of the molecule into separate atoms, the same is true for the lines in each row. Moreover, two rows v_1' and v_2' are shifted with respect to each other by a distance corresponding to the frequency difference of two vibrational levels v_1' and v_2' of the lower state. By superposition of all rows we obtain the whole system of vibrational transitions as indicated at the top of the figure, which at first sight does not show great regularity. From what has been said above it is apparent that the distances ν_{diss} and ν'_{diss} represent the frequencies corresponding to dissociation of the molecule from the lowest vibrational levels of the upper and lower electronic states respectively, and that the total length of the spectrum is $\nu_{\text{diss}} + \nu'_{\text{diss}}$.

Each of the lines in Fig. 9 represents in reality a *band* since on every vibrational level there is built up a set of rotational levels. If for the moment we confine our attention to singlet levels and disregard any fine structure due to Λ-doubling, we may approximately write according to Art. 5 for the energy of the rotational levels J and J' of given vibrational levels in the upper and lower electronic states:

$$W_J = W + B_0 \left(J + \tfrac{1}{2} \right)^2,$$

$$W_{J'} = W' + B_0' \left(J' + \tfrac{1}{2} \right)^2.$$

Since according to selection rule 7 J may change only by unity or zero, we obtain three kinds of transitions whose frequencies will be given by

$$h\nu_{JJ \pm 1} = W - W' - B_0' + (B_0 - B_0')\left(J + \tfrac{1}{2}\right)^2 \mp 2B_0'\left(J + \tfrac{1}{2}\right), \quad (4)$$

$$h\nu_{JJ} = W - W' + (B_0 - B_0')\left(J + \tfrac{1}{2}\right)^2. \ldots\ldots\ldots(5)$$

These frequencies are most conveniently represented by a diagram as shown in Fig. 10. Here the frequency is again the abscissa, while $(J + \frac{1}{2})$ is used as the ordinate. The relations (4), (5) give then three parabolas, of which only the parts above the frequency axis concern us, J being positive. The intersections of the lines $J = 0, 1, 2, \ldots$ with the parabolas are points, the abscissae of which represent the frequencies of the band lines as given by equations (4) and (5).

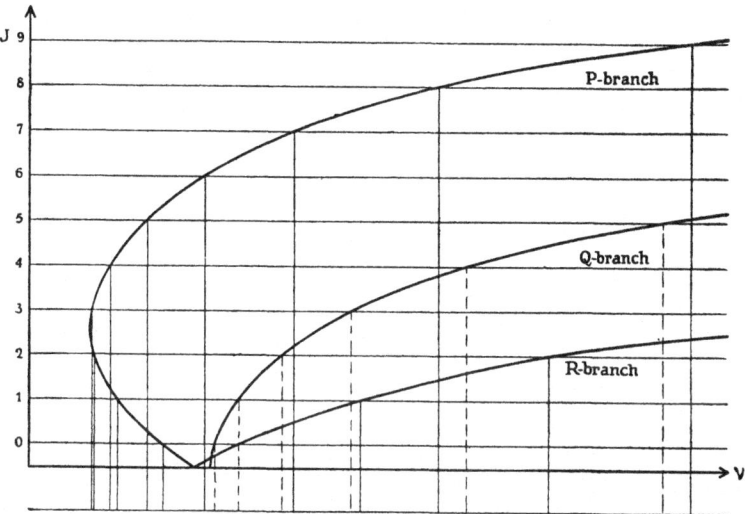

Fig. 10. Generation of the lines in a band from the transitions between the rotational levels J and $J' = J$ or $J \pm 1$ of given vibrational levels in two different electronic states.

Several circumstances ought to be noticed in connection with Fig. 10. From equation (4) it follows that the parabola called R-branch is the image of the continuation of the parabola called P-branch below the ν-axis when it is reflected at the ν-axis. Furthermore, the parabolas will be opened either toward the high or the low frequency side depending on whether in equation (4) $B_0 > B_0'$ or $< B_0'$, i.e. according to whether the moment of inertia of the molecule in the upper state is smaller or larger than that in the lower state. Finally, since the smallest value J can take is $|\Omega|$, J may take this value in the equation (4) with the upper sign,

while in that with the lower sign the smallest value is $|\Omega| + 1$.
Considering only the P- and R-branch, there hence will be $2|\Omega| + 1$
lines missing in the sequence of lines indicated in the figure (one
line for $\Omega = \Omega' = 0$). The place at which the lines are crowded
together due to the folding-over of the parabola P is called the
band head.

If we are dealing with multiple levels and if we consider the fine
structure due to Λ-doubling, we shall have instead of each parabola
in Fig. 10 several parabolas lying close together. Depending on

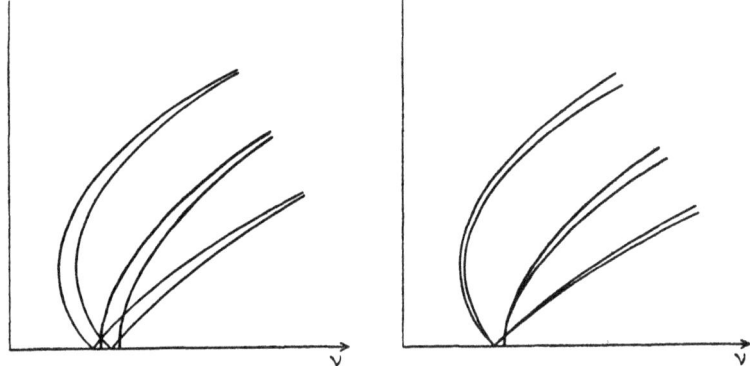

Fig. 11. Examples of multiplet and fine structure in the parabolas of Fig. 10.

whether we are dealing with the multiplet structure of case (a) or
with Λ- and ρ-doubling, the component parabolas will be separated
for small values of J and come closer together for increasing rotation,
or the reverse will happen. For according to Arts. 9 and 10 the
rotation diminishes the width of the spin multiplets, while it causes
the Λ- and ρ-fine structure. Fig. 11 illustrates qualitatively the
behaviour of the parabolas in the two cases.

There remains finally the question of the *intensities of the lines
in electronic bands*. The answer to this question is furnished by the
expression (1), Art. 14, in the case of emission bands and the
expression (2), Art. 14, in the case of absorption bands, together
with the considerations in sections 8 to 15 of this article regarding
the probability of emissive transitions and the relation (3), Art. 14,

connecting it with the probability of absorptive transitions. In the expressions (1), (2) of Art. 14 there enters the number of molecules in the different stationary states. In absorption experiments the gas will usually be in temperature equilibrium, and the distribution of the molecules over the different stationary states will be given by the law of Boltzmann. At ordinary temperatures in particular molecules will in general be present in appreciable numbers only in the rotational levels of the lowest vibrational level belonging to the normal electronic state. In neighbouring rotational levels whose energy difference is small compared to kT, where k is Boltzmann's constant and T the absolute temperature, the relative number of molecules will be practically given by the ratio of the statistical weights of the levels. These weights one would expect to be $2J + 1$, where J is the total angular momentum, if Λ- and ρ-doubling have removed all degeneracies except that of spatial orientation, since then the molecule has just $2J + 1$ possible orientations of its angular momentum with respect to an external field. Hence one would expect the intensities of the band lines corresponding to successive values of J to change quite regularly. This is indeed found to be the case except in homonuclear molecules for which a curious *alternation of intensities* takes place in many band systems, such that successive lines in a band are alternately strong and weak, the weak lines in some cases even being entirely absent. The cause of this important phenomenon will be considered in Art. 18.

In case a molecule is composed of atoms having *isotopes*, the spectra of the various isotopic molecules will not coincide exactly since, as we saw in Art. 5, the energy values are slightly different for them. A large amount of experimental material is available showing this effect. In the case of O and C new isotopes have actually been discovered from the spectral evidence, the relative proportion of which was so small that they had not been detected by the mass spectrograph. [See Birge and King (2), (3), (4), (37), (38).]

16. Vibrational Bands.

We proceed with the discussion of those transitions in diatomic molecules during which the electronic quantum numbers remain unaltered while the vibrational quantum number undergoes a change. These transitions possess frequencies lying in the infra-red and have only been observed in absorption. The experimental data refer hence exclusively to pairs of vibrational levels belonging to the normal electronic state, one of which is the lowest vibrational level of this state since in the higher ones no molecules are present at ordinary temperatures. We shall enumerate now the various selection rules and the formulae for the probabilities of transition applying to the type of band considered.

1. Rules 1 and 2 of Art. 15 are valid here just as in the case of electronic bands. *They lead for homonuclear molecules to the conclusion that vibrational transitions do not take place at all.* For we have seen in Art. 13 that all the even levels belonging to the same electronic state in a homonuclear diatomic molecule have the same kind of symmetry—are symmetrical, say—while all the odd levels have the opposite kind of symmetry. Thus, while rule 1 permits only transitions between even and odd levels, rule 2 just prohibits these transitions. The absence of vibrational bands in homonuclear molecules is a well-known experimental fact.

2. *As long as the nuclear vibration may be considered as simple harmonic, the vibrational quantum number v can change only by unity.*

This result, which follows directly if in the expression (5), Art. 14, we substitute for the vibrational part $P(\rho)$ of Ψ^0 the wave functions of the linear harmonic vibrator, loses its exact validity when the anharmonic terms in the nuclear binding become noticeable. Then v may change by more than unity, and indeed such "overtones" were found by Brinsmade and Kemble [265] in their early measurements of the vibrational bands of HCl. A discussion of the connection between the intensities of these overtones and the constants describing the anharmonic nature of the binding has been given by Fues (21) on the basis of wave mechanics.

3. In considering the rotational transitions belonging to a given vibrational transition we remember that all known diatomic gases, which have vibrational bands, possess as their normal state a $^1\Sigma$-state with the exception of NO, which has a $^2\Pi$-state. There is hence in no case ρ-doubling, and the Λ-doubling for NO will be too small to be resolvable in the infra-red. We are hence only concerned with the quantum number J which determines the total angular momentum of the molecule and *which may only change by unity or zero*. The relative transition probabilities are again given by the set of expressions in section 11 of Art. 15 corresponding to the case $\Omega' = \Omega$, from which it follows in particular that in molecules with a $^1\Sigma$-state as their normal state also the transitions $J' = J$ are absent.

The actual *appearance of vibrational bands* on spectrograms will be very different from that of the electronic bands. If there are "overtones," i.e. if the vibrational quantum number changes by more than one unit, it follows from the expression for the vibrational energy of Arts. 4 and 5 that their frequencies will be approximately integral multiples of the frequency of the fundamental band, for which v changes by unity. The frequencies of the different rotational lines belonging to the same vibrational band may be obtained just as in Art. 15, and we can use the equations (4), (5) there as approximate expressions for these frequencies if under $W - W'$ we now understand the difference in vibrational energy and if we put $B_0' = B_0$. For the moment of inertia will not be influenced very appreciably by a change in the vibrational energy alone without a change in the electronic configuration. We find thus

$$h\nu_{JJ\pm1} = W - W' - B_0 \mp 2B_0(J + \tfrac{1}{2}), \dots\dots\dots(1)$$

$$h\nu_{JJ} = W - W'. \dots\dots\dots\dots(2)$$

For molecules in a $^1\Sigma$-state the transitions in which J does not change, as mentioned before, have zero intensity so that there are just two branches of approximately equidistant lines, a P- and an R-branch. In the equation (1) with the upper sign J may have as smallest value zero, while in the equation (1) with the lower sign the smallest value of J is one. From this it follows that in the sequence of equidistant lines of the P- and R-branch there is one line missing. For NO there will also be a Q-branch corresponding to the transitions of equation (2). Since these all have the same frequency, the Q-branch will consist of a single line, which may be broadened, since on account of the approximate nature of equation (2) the coincidence of the frequencies is really not exact.

Again the *relative intensities* of the different rotational lines of the same vibrational band, which are here always observed in absorption, may be obtained by finding from the results for the probability of emission in section 3 the probability of absorption according to equation (3), Art. 14, and by assuming Boltzmann's law to furnish the distribution of the molecules over the different stationary states, the gas being in temperature equilibrium. As shown by experiments of Bourgin (263), (264) on the vibrational bands of HCl, the theoretical predictions are confirmed. It was in these bands, too, that an *isotope effect* was first observed and interpreted as such by Loomis (52) and Kratzer (44).

17. ROTATIONAL BANDS.

The last class of bands to be considered is that in which both the electronic and the vibrational quantum numbers remain unaltered and only the rotational quantum number changes, and which hence lie in the far infra-red. Here conditions are even simpler.

Rules 1 and 2 of Art. 15, which are quite general, lead just as in the case of vibrational bands to the conclusion that *rotational bands are not present in the spectra of homonuclear molecules.*

Also J *may change only by zero or unity*, and the probabilities of emissive transitions are found according to equation (4), Art. 14, from the following expressions for $\mathfrak{P}(j; j') \mathfrak{P}(j'; j)$:

$$J' = J - 1, \qquad |\mathfrak{P}|^2 \frac{J^2 - \Omega^2}{J(2J+1)},$$

given in the work of Mensing (61), Oppenheimer (78), Dennison (14), Kronig and Rabi (46), and Rademacher and Reiche (79). Here $|\mathfrak{P}|$ is the magnitude of the permanent dipole moment of the molecule so that not only the relative but also the absolute values of the transition probabilities are known. The transition probabilities in absorption may then be obtained from equation (3), Art. 14.

From the approximate expression for the rotational energy of the molecule of Art. 5,

$$W = W' + B_0 (J + \tfrac{1}{2})^2,$$

there results for the frequencies of the rotational band lines

$$\nu_{JJ-1} = 2B_0 J, \qquad \dots\dots\dots\dots\dots\dots\dots(1)$$

which expression, together with the conclusion regarding the transition probabilities, has been confirmed by the experimental work of Czerny (279), (280), and Badger (278) on the rotational bands of the hydrogen halides.

Some remarks may be made in this connection regarding the *rotational bands of polyatomic molecules*. If two of the principal moments of inertia are equal, the rotational energy is given by equation (1), Art. 7. Again *J may only change by zero or unity* while for reasons of symmetry Ω *will not be able to change*. It is then found that the frequencies of the rotational spectrum are given by an expression of the form (1). A molecule of the type considered is NH_3, and the measurements of Badger and Cartwright (283) have confirmed the theoretical expectation regarding the frequencies of the rotational lines.

18. BAND SPECTRA AND NUCLEAR STRUCTURE.

In Art. 15 it was mentioned that homonuclear diatomic molecules show the phenomenon of *intensity alternation in some of their electronic bands*. This phenomenon can be interpreted in all cases by assuming that the statistical weight of a level having no other degeneracies besides that of spatial orientation is not given by $2J + 1$ where J determines the total angular momentum of the molecule, but that the symmetrical levels have a weight $g_s (2J + 1)$ and the antisymmetrical ones a weight $g_s (2J + 1)$. If g_s or g_a is zero in particular, the band lines arising from transition between the symmetrical or the antisymmetrical levels will be absent entirely.

It remains now to show why the weights should be given by $g_s (2J + 1)$ and $g_a (2J + 1)$ rather than simply by $2J + 1$. We know from the spectral evidence (e.g. the He-spectrum) that for an atomic system containing two electrons only those states are realised in nature the wave functions of which are antisymmetrical in the coordinates (both positional and spin) of the two electrons, a result referred to already in Art. 3. If for atomic nuclei we postulate too that only one type of wave functions, either the symmetrical or the antisymmetrical, shall actually occur, and if the nuclei have no spin so that the positional coordinates suffice to determine them completely, then either the states called antisymmetrical or those called symmetrical, meaning in the positional coordinates of the nuclei, should be absent so that either g_a or g_s would be zero.

If, however, the *nuclei have a spin*, then a function antisymmetrical in the positional coordinates of the nuclei can be supplemented by multiplication with a function of the spins in such a way that the product is symmetrical when all the coordinates of the two nuclei are interchanged, and similarly a symmetrical function of the positional coordinates of the nuclei may be supplemented to become antisymmetrical. For the case of two electrons we have already illustrated this process in detail in Art. 3, showing that from one wave function symmetrical in the positional coordinates of the electrons there result three wave functions symmetrical and one antisymmetrical in all the coordinates of the electrons, while a function antisymmetrical in the positional coordinates of the electrons gives rise to one symmetrical and three antisymmetrical ones. Excluding the states with wave functions symmetrical in all the coordinates we see that here a state having originally a wave function antisymmetrical in the positional coordinates of the electrons gets a weight three times as great as a state having originally a wave function symmetrical in the positional coordinates of the electrons.

This result must be generalised in two ways for the case of nuclei. For in order to obtain agreement with the experimental evidence it is necessary in the first place to assume that the nuclear spin, instead of having a magnitude $\frac{1}{2}$ and two possible orientations, parallel and antiparallel, in an external magnetic field, in general has a magnitude s_n and $2s_n + 1$ possible orientations in an external magnetic field. s_n is one of the numbers $0, \frac{1}{2}, 1, \frac{3}{2}, \ldots$, but has a perfectly definite value for every nucleus. The different orientations of s_n are characterised by its projection on the direction of the magnetic field taking the values $-s_n, -s_n + 1, \ldots s_n$. In the second place, while in the case of electrons only the wave functions antisymmetrical in all the coordinates of the two electrons are realised in nature, it is found that for some nuclei the antisymmetrical wave functions and for others the symmetrical wave functions are the ones actually occurring. One expresses this briefly by saying that to the first kind of nuclei the *statistics of Dirac and Fermi*[†] applies, while the others are subject to the *statistics of Bose and Einstein*[‡].

Just as in the case of two electrons in Art. 3 with spin $\frac{1}{2}$ we constructed four spin functions, three of which were symmetrical

[†] Dirac, *Proc. Roy. Soc.* A **112**, 661, 1926; Fermi, *Zeit. f. Phys.* **36**, 902, 1926.
[‡] Bose, *Zeit. f. Phys.* **26**, 178, 1924; Einstein, *Sitz. preuss. Ak.* 1924, 261; 1925, 3, 18.

and one antisymmetrical, so in the more general case of a spin s_n we get $(2s_n + 1)^2$ spin functions of which $(s_n + 1)(2s_n + 1)$ are symmetrical and $s_n(2s_n + 1)$ antisymmetrical in the spin coordinates. If a state has a wave function symmetrical in the positional coordinates of the nuclei, this can be supplemented in $s_n(2s_n + 1)$ ways to become antisymmetrical in all the coordinates, while a state with a wave function antisymmetrical in the positional coordinates of the nuclei can be made to stay antisymmetrical in all the coordinates in $(s_n + 1)(2s_n + 1)$ ways. For nuclei, for which only wave functions antisymmetrical in all the coordinates are to be retained and which hence obey the statistics of Dirac and Fermi, we have thus:

$$\frac{g_s}{g_a} = \frac{s_n}{s_n + 1}. \qquad \ldots\ldots\ldots\ldots\ldots\ldots(1)$$

Similarly for nuclei obeying the statistics of Bose and Einstein

$$\frac{g_s}{g_a} = \frac{s_n + 1}{s_n}. \qquad \ldots\ldots\ldots\ldots\ldots\ldots(2)$$

From the intensity alternation in the band spectra of homonuclear molecules it is hence possible to determine the magnitude of the nuclear spin, as first pointed out by Heisenberg[23] and Hund[31]. The following table contains a summary of the experimental data available to-day on this topic.

Element	Intensity ratio	s_n	Statistics	References
H_1	$1:3$	$\frac{1}{2}$	DF	Kapuscinski and Eymers[99], specific heat (Art. 25)
He_4	$0:2$	0	BE	Half of lines missing
C_{12}	$0:2$	0	?	Half of lines missing
N_{14}	$2:4$	1	BE	Ornstein and Van Wijk[149]
O_{16}	$0:2$	0	BE	Half of lines missing
F_{19}	$1:3$	$\frac{1}{2}$	?	Gale and Monk[158]
Na_{23}	Alternation not pronounced	Large	?	Loomis and Wood[162]
Cl_{35}	$5:7?$	$\frac{5}{2}$?	?	Elliott[168]
I_{127}	Alternation not pronounced	Large	?	Loomis[173]

Column 1 gives the symbol of the element to be considered. Since several of the elements have isotopes, a subscript indicates to which isotope the measurements refer.

Column 2 contains the experimentally determined intensity ratios of the weak to the strong lines. If the weak lines are absent, we call this ratio for the sake of uniformity $0:2$. For Na_2 and I_2 there is no alternation as far as visual estimates of intensity go. It is to be expected that careful quantitative measurements will show a slight alternation. Finally for molecules the nuclei of which have isotopes an intensity alternation should occur only if both nuclei belong to one and the same isotope. Thus for Cl_2, bands arising from the molecules Cl_{35}—Cl_{35} and Cl_{37}—Cl_{37} should show alternation while the bands of Cl_{35}—Cl_{37} should not. Bands of Cl_{37}—Cl_{37} have thus far not been investigated since the proportion of Cl_{37} in chlorine gas is rather small, but for the other two kinds of chlorine molecules the theoretical expectation was confirmed by measurements of Elliott (168). Also the fact that the molecules O_{16}—O_{17}, O_{16}—O_{18}, and C_{12}—C_{13} do not have half of the lines missing like the molecules O_{16}—O_{16} and C_{12}—C_{12} led Birge and King (2), (3), (4), (37), (38) to the discovery of the isotopes O_{17}, O_{18} and C_{13} of oxygen and carbon.

Column 3 in the table shows the value of the nuclear spin as determined from the intensity ratio given in column 2 by means of the equations (1) or (2). In the case of H the value is confirmed by measurements of the specific heat of molecular hydrogen as we shall see in Art. 25. A rather striking result is obtained in the case of N, which was first pointed out by Kronig (50). Both protons and electrons have a spin of magnitude $\frac{1}{2}$. Now we know that for the electrons outside the nucleus the individual spins form a resultant which has one of the values $0, 1, 2, \ldots$ if the number of electrons is even, and one of the values $\frac{1}{2}, \frac{3}{2}, \frac{5}{2}, \ldots$ if the number of electrons is odd. One might expect the same to hold for the particles, protons and electrons, of which we consider the nuclei built up. However, since the N-nucleus contains an odd number of particles, as is apparent from the fact that its charge is an odd multiple of the elementary charge, the experimental result of a nuclear spin 1 is at variance with this expectation. We must hence conclude that in the process of forming a nucleus protons and electrons do not retain their individuality in the same way as they do in atom or molecule formation.

Column 4 in the table shows whether the nuclei are subject to the Dirac-Fermi (DF) statistics or the Bose-Einstein (BE) statistics. In order to decide this question on the basis of the experimental data it is necessary, according to equations (1) and (2), to know if the strong transitions are those between the symmetrical or those between the antisymmetrical levels. This question cannot in most cases be decided because, although for a given electronic state the subdivision into symmetrical and antisymmetrical levels is apparent, it is not known which of the two sets of levels is symmetrical and which antisymmetrical. This uncertainty arises from the fact that we usually do not know how the electronic part Φ of the approximate wave function Ψ^0 belonging to the electronic state in question behaves under the coordinate transformation representing an interchange of the two nuclei. However, if the electronic state is a $^1\Sigma$-state which arises if two equal atoms in the same S-state are brought together, we know according to Wigner and Witmer [85] that the rotational levels with even K are symmetrical, those with odd K antisymmetrical. This condition is fulfilled for the normal states of H_2 and N_2. For the former, as we shall see in Art. 25, it follows from the behaviour of the specific heat at low temperatures that the levels with odd K have the greater statistical weight. From the remarks just made we conclude that $g_a > g_s$ and that hence according to equations (1) and (2) the Dirac-Fermi statistics applies to protons. In the case of N_2 the measurements of Rasetti [295], [298], on the Raman effect of that gas, to be discussed in Art. 20, lead to the result that for the normal state of N_2 the rotational levels with even K have the greater statistical weight. Hence $g_s > g_a$ so that the Bose-Einstein statistics applies. This was first pointed out by Heitler and Herzberg (24) and is just as curious as the fact mentioned previously that the spin of the N-nucleus is an integer. For one would think that the interchange of two nuclei ought to be equivalent to the interchange in pairs of the particles, protons and electrons, of which they are composed. Now since each N-nucleus contains an odd number of particles, an odd number of interchanges of protons or electrons will be required to bring about a complete exchange of the N-nuclei, and at every interchange the wave function changes sign since both to the protons and to the electrons the Dirac-Fermi statistics applies. Hence when the N-nuclei are completely interchanged, one would expect the resulting wave function to have the sign opposite to that before the exchange and hence the Dirac-Fermi statistics instead of the Bose-Einstein statistics. We see thus that in the interior of nuclei new laws must come into action, and it is therefore all the more desirable to ascertain from band spectra for as many nuclei as possible their spin and statistical properties.

19. Transitions in the Stark and Zeeman Effect.

We have seen in Art. 6 that in general an electric or a magnetic field causes the energy levels of a diatomic molecule to be resolved into component levels. In consequence the spectral lines arising from the originally single levels will then be decomposed into a number of component lines. We wish now to investigate the *selection rules* limiting the possibilities of transitions between the component levels and the *relative probabilities of* those *transitions* permitted by them. In the discussion we shall limit ourselves to the Stark and Zeeman effect of such lines for which the initial and final levels both come under case (a) or both under case (b).

Considering first two levels both coming under case (a) and having angular momenta J and J' respectively, we shall have for the first according to Art. 6 $2J + 1$ component levels distinguished by the quantum number M and for the second $2J' + 1$ component levels distinguished by the quantum number M'. In order that combinations may take place at all, J' must not differ from J by more than unity. Furthermore M is subject to the selection rule that *M' may not differ from M by more than unity* for the transition probability to be different from zero. If the spectrum is observed at right angles to the direction of the field, the *component lines arising from transitions in which M changes by unity are linearly polarised perpendicular to the direction of the field, those arising from transitions in which M remains unchanged are linearly polarised parallel to the direction of the field.*

This last result is a consequence of the fact that if we calculate by means of equation (5), Art. 14, the x-, y-, and z-components $\mathfrak{P}_x(j;j')$, $\mathfrak{P}_y(j;j')$ and $\mathfrak{P}_z(j;j')$ of the matrix $\mathfrak{P}(j;j')$, representing the dipole moment of the molecule, then if the external field is in the direction z and if the states j and j' have the same M, only $\mathfrak{P}_z$ is different from zero while if they have an M differing by unity, only $\mathfrak{P}_x$ and $\mathfrak{P}_y$ will be different from zero. In this calculation we simply have to substitute for Ψ_j and $\Psi_{j'}$ under the integral sign of equation (5), Art. 14, the wave functions discussed in Chap. II.

The quantities $\mathfrak{P}(j;j')\,\mathfrak{P}(j';j)$ of equation (4), Art. 14, determined in this way are given by the following expressions, which are the same as those governing the intensity distribution among the component lines in the Stark and Zeeman effect of atomic spectra:

for $J' = J+1$, $M' = M \pm 1$ by

$$\text{const. } O(J+1, \pm M+1)/2(J+1)(2J+3);$$
$$M' = M,$$
$$\text{const. } Q(J+1, M)/(J+1)(2J+3);$$
$$J' = J, \quad M' = M \pm 1,$$
$$\text{const. } P(J+1, \pm M)/2J(J+1);$$
$$M' = M,$$
$$\text{const. } Q(0, M)/J(J+1);$$
$$J' = J-1, \quad M' = M \pm 1,$$
$$\text{const. } O(J, \mp M)/2J(2J-1);$$
$$M' = M,$$
$$\text{const. } Q(J, M)/J(2J-1),$$

where the functions O, P, Q are defined by equations (1), (2), (3), Art. 15. Here in each set of three expressions for a given change of J the constant is common to all transitions belonging to the set, i.e. independent of M. Addition of the three expressions gives in each case simply this constant (summation rule). The relative intensities of the component lines for observation at right angles to the field are given directly by the above expressions after those corresponding to a change of M by unity, i.e. to the perpendicular components, have been multiplied by $\frac{1}{2}$. All these results are not characteristic of molecules but according to quantum mechanics apply quite generally to an atomic system with angular momentum J in an external field.

If the two levels considered both come under case (b) and if the angular momenta of the molecule exclusive of the spin are K and K' for them respectively, then the component levels will be characterised by the quantum number M_K or M'_K specifying the projection of K and K' on the direction of the external field. All that has been said in the preceding pages may be taken over literally if only we replace J and J' by K and K', M and M' by M_K and M'_K.

For the previous considerations and particularly for the expressions determining the relative intensities of the component lines in the Stark and Zeeman effect to be valid, a number of conditions must be satisfied. For the levels coming under case (a) the interaction energy of the molecule with the outside field should be small compared both to the interaction energy by which the resultant electron spin is coupled to the molecule and to the energy difference between neighbouring rotational levels of the molecule. For the levels coming under case (b) on the other hand the interaction energy of the field should be large compared to the interaction energy of the spin and small compared to the energy difference between neighbouring rotational levels of the molecule.

A study of the Zeeman effect has been made by Kemble, Mulliken, and Crawford (258) and by Crawford (256) for bands of CO arising from a transition between a $^1\Pi$- and a $^1\Sigma$-level. They find good agreement with the theory both as regards the displacements of the component lines calculated from the energy changes due to the field, which we discussed in Art. 6, and their relative intensities discussed in this article. A systematic deviation of the intensities in some cases from those expected according to the expressions given was explained by the author (49) as due to the circumstance that the interaction energy of the molecule with the outside field was no longer small compared to the energy difference between neighbouring rotational levels.

CHAPTER IV

MACROSCOPIC PROPERTIES OF MOLECULAR GASES

20. SCATTERING.

The band spectra which have been investigated in Chaps. I, II, III may be said to be characteristic of the individual molecules of which a gas is composed and to reveal their internal structure. We come now to the discussion of properties such as are shown by molecular gases in bulk, a complete understanding of which has only been attained with the help of the detailed information regarding the structure of a single molecule derived from its band spectra.

We commence by studying the *behaviour of a gas under the influence of incident monochromatic radiation* of frequency ν. The quantum theory leads to the result that every molecule acts as a centre of secondary spherical wavelets having partially the same frequency as the incident radiation (*coherent scattering*) and partially modified frequencies (*incoherent scattering*).

If for the moment we suppose the frequency ν to lie in the optical or infra-red region of the spectrum so that the wave-length of the radiation may be taken as large compared to the molecular dimensions, then the scattered radiation is most conveniently described in terms of the dipole moments which according to classical electrodynamics would give rise to its emission. If the electric field of the incident radiation at the place where a molecule is situated is given by the real part of

$$\mathfrak{E}e^{2\pi i \nu t}, \quad \dots\dots\dots\dots\dots\dots\dots\dots(1)$$

and if the molecule is in the state j with energy W_j, then, as shown by Kramers and Heisenberg† under the assumption regarding the

† Kramers and Heisenberg, *Zeit. f. Phys.* **31**, 681, 1925.

wave-length made above, the total radiation scattered by it may be derived from the following sum of dipole moments:

$$\sum_{j'}^{W_{j'} < W_j + h\nu} [\mathfrak{p}(j;j') + \mathfrak{p}^*(j;j')] + \sum_{j'}^{W_{j'} < W_j - h\nu} [\mathfrak{p}(j';j) + \mathfrak{p}^*(j';j)].$$

$$.........(2)$$

Here j' is any state of the molecule whose energy $W_{j'}$ obeys the restrictions imposed at the top of the summation signs, the asterisk denotes the conjugate, and the quantity $\mathfrak{p}(j;j')$ is given by

$$\mathfrak{p}(j;j') = -\frac{1}{2h}\sum_{j''}\left[\frac{(\mathfrak{E}\mathfrak{P}(j;j''))\,\mathfrak{P}(j'';j')}{\nu_{jj''} + \nu}\right.$$

$$\left. + \frac{\mathfrak{P}(j;j'')(\mathfrak{E}\mathfrak{P}(j'';j'))}{\nu_{j'j''} - \nu}\right]e^{2\pi i(\nu_{jj'} + \nu)t},......(3)$$

where j'' is any state of the molecule different from j and j', and $\nu_{jj''}$, $\nu_{j'j''}$ are determined by equations analogous to equation (1), Art. 1, $\mathfrak{P}(j;j'')$, $\mathfrak{P}(j'',j')$ by equations analogous to equation (5), Art. 14.

In the first sum of equation (2) j' may be equal to j, and the corresponding term gives rise to the coherent scattered radiation since according to equation (3) $\mathfrak{p}(j;j)$ depends upon the time through the factor $e^{2\pi i\nu t}$. Without loss of generality we may assume that only the z-component $\mathfrak{E}_z$ of $\mathfrak{E}$ is different from zero so that the incident radiation is linearly polarised in the direction z. Now according to equation (3) $\mathfrak{p}(j;j)$ is composed of contributions due to the various transitions jj''. In order that for a given j'' the contribution may be different from zero, it is evidently necessary for $\mathfrak{P}(j;j'')$ or $\mathfrak{P}(j'';j)$ to have a z-component which does not vanish since otherwise the scalar products $(\mathfrak{E}\mathfrak{P}(j;j''))$ and $(\mathfrak{E}\mathfrak{P}(j'';j))$ in equation (3) will both be zero. For the application of equation (2) it is presupposed that the states j, j', j'' are non-degenerate. In fact there will in general be present the degeneracy of spatial orientation, but this we may imagine removed by a very small magnetic field parallel to the direction z. If we introduce in place of the one index j three indices q, J, M to characterise a stationary state of the molecule, one of them, M, may be chosen identical with the magnetic quantum number M which we have been using already in earlier chapters and which measured the angular momentum of the molecule in the direction of the outside field, while the other two, q and J, which will later be given a meaning somewhat different from that previously assigned to these letters, stand for the remaining quantum numbers required to specify the state of the molecule. Then according to the results of Art. 19 the two states j and j'' must have the same M if $\mathfrak{P}_z(j; j'')$ is to be different from zero, and for two such states both $\mathfrak{P}_x(j; j'')$ and $\mathfrak{P}_y(j; j'')$ vanish. Equation (3) gives therefore

$$\mathfrak{p}_z(q, J, M; q, J, M)$$
$$= -\frac{\mathfrak{E}_z}{h} \sum_{q'J'} \frac{\nu(q,J; q',J')\,\mathfrak{P}_z(q,J,M; q',J',M)\,\mathfrak{P}_z(q',J',M; q,J,M)}{\nu^2(q,J; q',J') - \nu^2} e^{2\pi i \nu t}$$
$$\dots\dots(4)$$

if we introduce single-primed letters in place of double-primed ones and remember that $\nu_{jj'}$ does not depend upon M since the splitting produced in the energy levels by the auxiliary magnetic field is made to disappear in the limit. $\mathfrak{p}_x$ and $\mathfrak{p}_y$ on the other hand both vanish according to equation (3) and the remarks made above. It follows then that *if the incident radiation is linearly polarised, the coherent scattered radiation will also be linearly polarised in a plane through the direction of polarisation of the incident radiation.*

Making use of equation (4), Art. 14, we see that the scattering moment $\mathfrak{p}(j; j)$ determining the coherent scattered radiation is intimately connected with the transition probabilities. In fact according to that equation

$$\mathfrak{P}_z(q, J, M; q', J', M)\,\mathfrak{P}_z(q', J', M; q, J, M)$$
$$= \frac{3c^3 h}{64\pi^4 \nu^3(q,J; q',J')} A^{qJM}_{q'J'M} \text{ if } W_{qJ} > W_{q'J'} \qquad \dots\dots(5)$$
$$\text{or} \qquad = \frac{3c^3 h}{64\pi^4 \nu^3(q',J'; q,J)} A^{q'J'M}_{qJM} \text{ if } W_{q'J'} > W_{qJ}.$$

We must now remember that if we know a molecule to be in a

state q, J, then on account of the spatial degeneracy we do not know yet what value M has. In order to obtain the scattering moment $\mathfrak{p}(q, J; q, J)$ of the molecule in the degenerate state q, J, we have therefore to average over all values of M. Equations (5) become thereby

$$\mathfrak{P}_z(q, J; q', J')\,\mathfrak{P}_z(q', J'; q, J)$$

$$= \frac{c^3 h}{64\pi^4 \nu^3(q, J; q', J')} A_{q'J'}^{qJ} \quad \text{if } W_{qJ} > W_{q'J'}$$

$$\text{or} \qquad = \frac{c^3 h}{64\pi^4 \nu^3(q', J'; q, J)} \frac{g_{q'J'}}{g_{qJ}} A_{qJ}^{q'J'} \text{if } W_{q'J'} > W_{qJ},$$

$$\qquad\qquad\qquad\qquad\qquad\qquad\qquad\qquad\qquad\qquad \dots(6)$$

where $A_{q'J'}^{qJ}$ is the transition probability from the state q, J to a lower state q', J', $A_{qJ}^{q'J'}$ the transition probability from a higher state q', J' to the state q, J, and where g_{qJ}, $g_{q'J'}$ are the statistical weights of the states q, J and q', J' respectively. In the derivation of equations (6) from equations (5) use is made of the fact that the transitions from the component levels M of the state q, J to those of the state q', J' if q', J' lies lower, or from the component levels M of the state q', J' to those of the state q, J if q', J' lies higher, in which M remains unchanged form one-third of all the transitions in which no restriction is imposed on the change of M.

If we introduce the abbreviations

$$f(q, J; q', J') = -\frac{c^3 m}{8\pi^2 e^2 \nu^2(q, J; q', J')} A_{q'J'}^{qJ} \quad \text{if } W_{qJ} > W_{q'J'},$$

$$f(q, J; q', J') = \frac{c^3 m}{8\pi^2 e^2 \nu^2(q', J'; q, J)} \frac{g_{q'J'}}{g_{qJ}} A_{qJ}^{q'J'} \quad \text{if } W_{q'J'} > W_{qJ},$$

$$\qquad\qquad\qquad\qquad\qquad\qquad\qquad\qquad\qquad\qquad \dots\dots\dots(7)$$

where e and m are charge and mass of an electron respectively, the averaging of equation (4) over M gives us according to equations (6):

$$\mathfrak{p}_z(q, J; q, J)$$

$$= -\frac{\mathfrak{E}_z}{h} \sum_{q'J'} \frac{\nu(q, J; q', J')\,\mathfrak{P}_z(q, J; q', J')\,\mathfrak{P}_z(q', J'; q, J)}{\nu^2(q, J; q', J') - \nu^2} e^{2\pi i \nu t}$$

$$= \mathfrak{E}_z \sum_{q'J'} \frac{e^2}{8\pi^2 m} \frac{f(q, J; q', J')}{\nu^2(q, J; q', J') - \nu^2} e^{2\pi i \nu t} \dots\dots\dots\dots\dots\dots(8)$$

Since the coherent scattering is determined by the dipole moment $\mathfrak{p}_z(q, J; q, J) + \mathfrak{p}_z^*(q, J; q, J)$, the molecule behaves, as far as this is concerned, like a system of charged particles on the classical theory elastically bound to an equilibrium position with frequencies $\nu(q, J; q', J')$ and having for the ratio of the square of their charge to their mass the value $f(q, J; q', J') e^2/m$. According to equations (7) emissive transitions starting from the state q, J give negative f-values, absorptive transitions positive f-values.

The complete polarisation claimed by the theory for the coherent scattered radiation if the incident radiation is polarised has been experimentally verified for gases at pressures sufficiently low so that the impacts between the gas molecules do not disturb the phenomenon.

From equations (2) and (3) we see that the radiation scattered by a molecule in the state j may contain besides the unmodified frequency ν the frequencies $\nu_{jj'} + \nu$ and $\nu_{jj'} - \nu$, the former only provided $\nu > \nu_{j'j}$, the latter provided $\nu_{jj'} > \nu$. Since scattering experiments are made with gases at ordinary temperatures and radiation in the optical or ultra-violet region of the spectrum, there will not be any molecules present with an emission frequency $\nu_{jj'}$ exceeding ν so that scattered radiation with frequencies of the type $\nu_{jj'} - \nu$ will not occur in actual practice. The frequencies $\nu_{jj'} + \nu$ can be smaller or larger than ν, depending on whether the state j' lies above or below the state j.

In order that a frequency $\nu_{jj'} + \nu$ *should* actually occur, it is necessary that the scattering moment $\mathfrak{p}(j; j')$ be different from zero. According to equation (3) this will only be the case if there exists at least one third state j'' such that both $\mathfrak{P}(j; j'')$ and $\mathfrak{P}(j'', j')$ do not vanish, i.e. a state combining both with j and j'.

Following the work of Raman on the scattering of light by liquids, Wood (303), (304) and Rasetti (294)–(299) have recently been able to show experimentally the presence of modified frequencies in the radiation scattered by molecular gases. The connection of this *Raman effect* with the quantum theory of scattering has been pointed out by a number of authors [Smekal (300), Born (284), Langer (292), Dieke (287), Rasetti (297), (298), Hill and Kemble (291)]. Some of these have also investigated in greater detail the question as to which modified frequencies will occur for different kinds of diatomic gases, a problem to which we now turn our attention.

The known diatomic gases with the exception of O_2 and NO have molecules which at ordinary temperatures will be practically all in the various rotational levels J of the lowest vibrational level be-

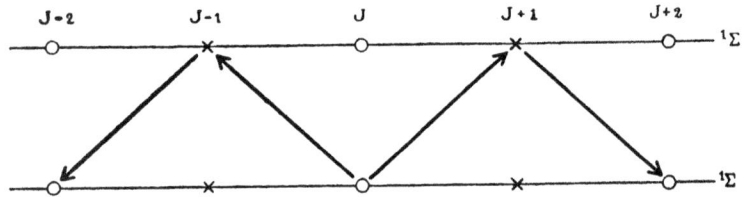

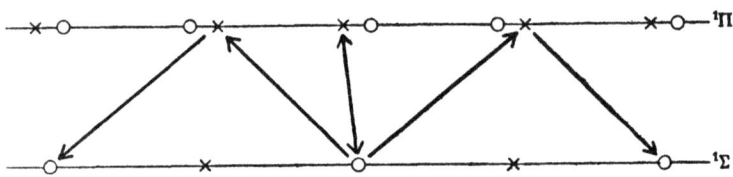

Fig. 12. Diagram showing transitions by which rotational Raman lines are produced for a diatomic molecule in a $^1\Sigma$-state.

longing to a $^1\Sigma$-state. From a particular level J, according to selection rules 3, 4 and 7 and the intensity formulae of Art. 15, there are transitions possible to the levels $J \pm 1$ of other $^1\Sigma$-states and to the levels $J, J \pm 1$ of $^1\Pi$-states. Moreover, according to selection rule 1, Art. 15, these last levels must be odd if the given level J is even and vice versa. By the same selection rules there are transitions possible from the intermediate levels back to the original level J and to levels $J \pm 2$, but to no other rotational levels of the lowest vibrational level of the normal $^1\Sigma$-state. This state of affairs is illustrated in Fig. 12 by separate diagrams for the transitions

$^1\Sigma$-$^1\Sigma$-$^1\Sigma$ and $^1\Sigma$-$^1\Pi$-$^1\Sigma$, successive rotational levels being placed on a horizontal with a cross for even states and a circle for odd ones. There should then appear in the Raman spectrum, besides the unmodified line, lines shifted by an amount $\pm\,\nu_{jj'}$ equal to the frequency difference between the rotational levels with values of J differing by two units. If, according to equation (2), Art. 5, we use as approximate expression for the rotational energy

$$B_0\,(J+\tfrac{1}{2})^2,$$

then the frequency of the shifted lines will be given by

$$\nu \pm 2B_0\,(2J+3), \quad J = 0, 1, 2, \ldots \quad \ldots\ldots\ldots\ldots(9)$$

Making use of selection rule 6 instead of 7, Art. 15, and substituting K for J, one concludes in a similar fashion for O_2, whose normal state is a $^3\Sigma$-state, that if one neglects the fine structure and assumes that transitions to $^3\Pi$-states are unimportant compared to transitions to other $^3\Sigma$-states (which probably is justified), only lines shifted by an amount $\pm\,\nu_{jj'}$ equal to the frequency difference between two rotational levels with values of K differing by two units should occur. In analogy wtih equation (9) their frequency will be given by

$$\nu \pm 2B_0\,(2K+3), \quad K = 0, 1, 2, \ldots .$$

For NO on the other hand, whose normal state is a $^2\Pi$-state, the selection rules permit frequency shifts corresponding to a change of J by one or two units.

The shifted lines just described may be said to form the *rotational Raman spectrum* of the gas. They have been observed for HCl by Wood (304) and for H_2, N_2, O_2 by Rasetti (295), (298), (299), who verified the conclusions reached above. From the intensity alternations of the shifted lines Rasetti was also enabled to conclude that while in the normal $^1\Sigma$-state of H_2 the odd rotational levels have the greater statistical weight, in the normal $^1\Sigma$-state of N_2 this applies to the even levels, a result which we have used in advance in Art. 18 to make an important deduction regarding the statistics of nuclei. Incidentally he could obtain the moment of inertia of the

N_2-molecule in its normal state from the spacing of the shifted lines, a quantity which had hitherto not been measured since the bands connecting the normal state with other states are so ultra-violet that it is difficult to resolve their rotational structure.

In molecules having a $^1\Sigma$-state as their normal state there can occur, besides the transitions from the intermediate levels to the rotational levels J, $J \pm 2$ discussed above, other transitions ending

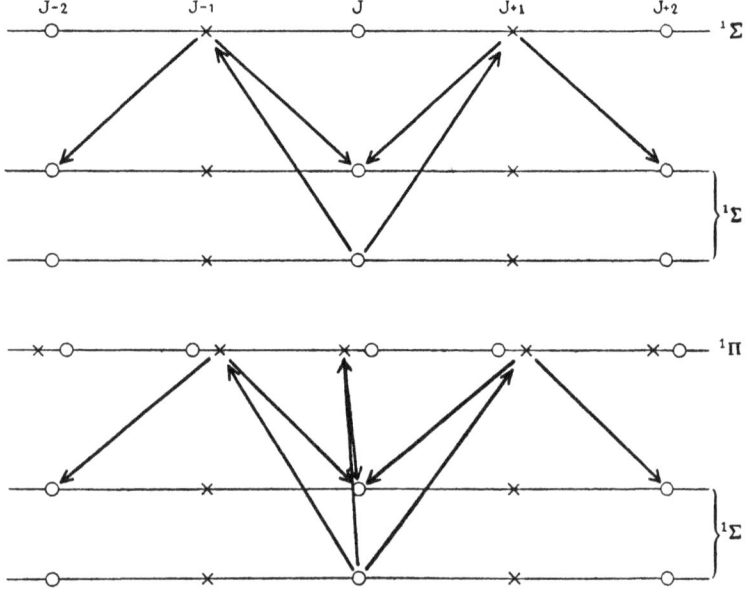

Fig. 13. Diagram showing transitions by which vibrational Raman lines are produced for a diatomic molecule in a $^1\Sigma$-state.

on the rotational levels J, $J \pm 2$ of a higher vibrational level belonging to the normal $^1\Sigma$-state. This behaviour is illustrated in Fig. 13 by two separate diagrams, analogous to those of Fig. 12, for the case where the intermediate levels belong to a $^1\Sigma$-state and for the case where they belong to a $^1\Pi$-state. Those frequencies $\nu_{jj'}$ which correspond to a change of vibrational quantum number only, while J remains unchanged, all coincide, no matter what the value of J is, and hence give in the Raman spectrum a strong line shifted by the amount of the frequency difference ν_0 of the two vibrational levels. Those corresponding also to a change of J by ± 2 give a set of lines on each side of the shifted line just spoken of with frequencies

$$\nu - \nu_0 \pm 2B_0\,(2J+3), \quad J = 0, 1, 2, \ldots.$$

The whole group of lines may be called the *vibrational Raman spectrum* of the gas and has been studied for HCl by Wood (303), (304) and for N_2, O_2, CO by Rasetti (294), (297). Quite similar remarks as previously apply here to O_2 and NO. In the Raman spectrum of the latter Rasetti (296), (299) has also been able to find a line shifted by the frequency difference of the states $^2\Pi_{\frac{1}{2}}$ and $^2\Pi_{\frac{3}{2}}$. As pointed out by Manneback (293) and Van Vleck (302) vibrational transitions in which v changes by more than unity should not occur in the Raman spectrum.

We have seen in the preceding paragraphs how the Raman spectrum of diatomic molecules is connected with their rotational and vibrational properties. It may thus be used to obtain information about these properties, a remark particularly important when applied to polyatomic molecules regarding which our knowledge is at present very incomplete. Some experimental work along that line has been done by Wood (304) and by Dickinson, Dillon, and Rasetti (288).

We shall now drop the assumption made at the beginning of this article that the wave-length of the radiation is large compared to the molecular dimensions and say a few words about the *scattering phenomena in the X-ray region*. They can be understood if it be remembered that there the electric vector of the incident radiation has no longer the same phase in different parts of the molecule. As a rough model of a diatomic molecule which, however, suffices to illustrate the essential features of the phenomena occurring we can take a system of two equal scattering atoms at a fixed distance l apart, and we shall study the scattering of an ensemble of such systems in which the line joining the atoms may have all directions in space with equal probability. According to Ehrenfest (289) the scattered intensity in a direction which makes an angle ϕ with that of the incident rays is given by

$$I = I_0 \left(1 + \frac{\sin 2\pi\tau}{2\pi\tau}\right), \quad \ldots\ldots\ldots\ldots\ldots(10)$$

where

$$\tau = \frac{2l}{\lambda}\sin\frac{\phi}{2} \quad \ldots\ldots\ldots\ldots\ldots\ldots(11)$$

and where I_0 denotes the intensity of the scattered radiation which would be obtained in the direction ϕ if all the atoms were independent instead of being coupled in pairs to form molecules.

In equation (10) not only the factor in the parenthesis but also I_0 will depend upon ϕ. I_0, however, which describes the scattering of single atoms, should theoretically, and does according to experiment, decrease uniformly with ϕ from its maximum at $\phi = 0$ to smaller values as long as ϕ does not go beyond 90°. The factor in the parenthesis on the other hand is an oscillating function which introduces subsidiary maxima and minima into the scattering curve. For different values of the ratio l/λ the following table taken from Ehrenfest's paper gives the angles for which the consecutive maxima and minima occur.

l/λ	ϕ_0	ϕ_1	ϕ_2	ϕ_3
$\frac{1}{2}$	0°	90°	—	—
1	0	41	71°	114°
2	0	21	34	50
3	0	14	22	32

Such maxima and minima were actually found for a number of molecular gases by Debye and his pupils(286) and have been utilised by them to determine the geometrical arrangement of the atoms in the molecules composed of them.

The proof of equation (10) is conducted as follows. Calling the two atoms in the molecule 1 and 2, the electric vector of the scattered radiation due to 1 at a distant point lying in the direction ϕ from the molecule will be given by $\epsilon \cos 2\pi\nu t$, that due to 2 by

$\mathfrak{e} \cos 2\pi \left(\nu t - \dfrac{\Delta}{\lambda}\right)$, where Δ is the difference in optical path of the two rays from 1 and 2. The total electric vector of the scattered radiation will be the sum, and the total scattered intensity will be proportional to its square, which gives

$$I = \text{const. } \mathfrak{e}^2 \left(1 + \cos 2\pi \frac{\Delta}{\lambda}\right). \quad \ldots\ldots\ldots\ldots(12)$$

Atom 2 may lie with its centre with equal probability anywhere on a sphere of radius l around the centre of atom 1. The expression (12) has therefore still to be averaged over all these positions. The result of this process is

$$I = \text{const. } \mathfrak{e}^2 \left(1 + \frac{\sin 2\pi\tau}{2\pi\tau}\right), \quad \ldots\ldots\ldots\ldots(13)$$

where τ is defined by equation (11). On the other hand if the same atoms were not united in pairs but formed a monatomic gas, the scattering would be simply

$$I_0 = \text{const. } \mathfrak{e}^2 \quad \ldots\ldots\ldots\ldots\ldots\ldots(14)$$

with the same constant as above. From equations (13) and (14) equation (10) follows.

21. DISPERSION.

The coherent scattered radiation described in the preceding section gives rise to the *refraction* of the incident radiation in the gas which it traverses, for the individual coherent spherical wavelets coming from the molecules will be superimposed in the forward direction to form a plane wave of the same frequency as the incident radiation, which by interference then gives rise to the retardation or acceleration of phase of which the phenomenon of refraction consists.

If the gas is composed of one sort of molecules only, and if there are N_{qJ} of them per unit volume in the stationary state qJ, then the term in the electric polarisation per unit volume which vibrates with frequency ν will be according to the preceding article

$$\sum_{qJ} N_{qJ} [\mathfrak{p}_z(q, J; q, J) + \mathfrak{p}_z{}^*(q, J; q, J)],$$

provided the electric field of the incident radiation is again given by the real part of the expression (1), Art. 20. According to Maxwell's theory the index of refraction n is obtained by putting $n^2 - 1$ equal to 4π times the ratio of this polarisation to the electric field of the incident light. Using equation (8), Art. 20, we find thus

$$n^2 - 1 = -\frac{8\pi}{h} \sum_{qJ} N_{qJ} \sum_{q'J'} \frac{\nu(q, J; q', J')\, \mathfrak{P}_z(q, J; q', J')\, \mathfrak{P}_z(q', J'; q, J)}{\nu^2(q, J; q', J') - \nu^2}.$$
.........(1)

The distribution of the molecules over the various stationary states q, J in a gas in temperature equilibrium is given by the law of Boltzmann, according to which

$$N_{qJ} = \frac{N_0}{Z}\, g_{qJ} e^{-\frac{W_{qJ}}{kT}}, \quad \dots\dots\dots\dots(2)$$

where N_0 is the total number of molecules per unit volume, g_{qJ} and W_{qJ} the statistical weight and the energy of the state qJ, k Boltzmann's constant, given by

$$k = 1371.10^{-16} \text{ erg/degree}, \quad \dots\dots\dots\dots(3)$$

T the absolute temperature, and

$$Z = \sum_{qJ} g_{qJ} e^{-\frac{W_{qJ}}{kT}}. \quad \dots\dots\dots\dots\dots(4)$$

We shall now specify the significance of the indices q, J more accurately. If in the molecule we imagine the multiplet, vibrational, rotational, and fine structure reduced to zero, we have only the different electronic levels left, which we shall distinguish by the index q. In all molecules the energy difference between the first excited and the normal electronic level is so large compared to kT at ordinary temperatures that according to equation (2) there will be practically no molecules present in this and the higher excited states, and in consequence in the quantity Z defined by equation (4) as well as in equation (1) the summation over q may be omitted, q taking only the one value q_0 of the normal state. The various multiplet, vibrational, rotational, and fine structure levels we shall then specify by the index J. The levels J belonging to the same value of q will lie much closer together than the groups of levels J belonging to different values of q, so that at ordinary temperatures molecules may be present in different levels J of the normal state q_0.

From the circumstances just described two important conclusions may be drawn concerning equation (1). On account of the factor $\nu(q, J; q', J')$ in the numerator the terms with $q' = q = q_0$ will be negligible compared to those for which $q' \neq q_0$ if ν lies in the optical region of the spectrum, since $\nu(q_0, J; q_0, J')$ is a vibrational or rotational frequency of the molecule and hence small compared to ν. In the summation over q' this index is hence to be given only values different from q_0. For $q' \neq q_0$, $\nu(q_0, J; q', J')$ will not depend much on J and J' and may hence be put approximately equal to $\nu(q_0; q')$, the frequency difference of the electronic states q' and q_0 whose multiplet, vibrational, rotational, and fine structure has been reduced to zero. Equation (1) may now be written

$$n^2 - 1 = -\frac{8\pi}{h} \sum_J N_{q_0 J} \sum_{q' \neq q_0} \frac{\nu(q_0; q')}{\nu^2(q_0; q') - \nu^2}$$
$$\times \sum_{J'} \mathfrak{P}_z(q_0, J; q', J') \, \mathfrak{P}_z(q', J'; q_0, J). \quad \ldots(5)$$

A further simplification results from the fact that the last sum in equation (5) determines, according to equations (6), Art. 20, and equation (3), Art. 14, the total probability of a transition from the state q_0, J to any of the levels J' of the state q. This probability is practically independent of J, the slow frequencies of vibration or rotation having little influence on the total amount of radiation emitted or absorbed (summation rule). We may hence replace the last sum by $\mathfrak{P}_z(q_0; q') \mathfrak{P}_z(q'; q_0)$, which permits us to perform the first summation, $\sum_J N_{q_0 J}$ being simply equal to N_0. Equation (5) thus becomes

$$n^2 - 1 = -\frac{8\pi N_0}{h} \sum_{q'} \frac{\nu(q_0; q') \, \mathfrak{P}_z(q_0; q') \, \mathfrak{P}_z(q'; q_0)}{\nu^2(q_0; q') - \nu^2}. \quad \ldots(6)$$

From this formula we conclude: *For frequencies ν lying in the optical region of the spectrum, the quantity $n^2 - 1$ is proportional to* *the density of the gas and independent of the temperature.* These results, which were first proved by Van Vleck (322) and have been confirmed by experiment, cannot be expected to hold when the frequency ν lies in the immediate neighbourhood of a natural frequency of a molecule of the gas, for then it will not be permissible any more to substitute $\nu(q_0; q')$ for $\nu(q_0, J; q', J')$ in equation (1). The general way in which $n^2 - 1$ depends upon the frequency ν has also been confirmed by experiment.

We want finally to make some remarks here regarding the *determination of the number of molecules per unit volume of a gas* from scattering experiments.

As we have just seen, the coherent scattering moment is practically the same for all the stationary states in which molecules are found to an appreciable extent at ordinary temperatures and may simply be called $\mathfrak{p}_z$. This equality of the scattering moment for all the molecules is presupposed for the validity of the relation first derived by Rayleigh, viz.

$$\frac{I}{I_0} = \frac{8\pi^3 (n^2 - 1)^2}{3\lambda^4 N_0}, \quad \dots\dots\dots\dots\dots(7)$$

where I is the energy of the coherent radiation scattered per second by unit volume of the gas containing N_0 molecules if the incident radiation, of wave-length λ, has an intensity I_0. For then according to electrodynamics

$$I = N_0 \frac{2}{3c^3} 16\pi^4 \nu^4 \overline{\mathfrak{p}_z^2},$$

the dash denoting the time average, while

$$n^2 - 1 = \frac{4\pi N_0 \mathfrak{p}_z}{\mathfrak{E}_z}.$$

Eliminating $\mathfrak{p}_z$ we get

$$I = \frac{2\pi^2 \nu^4 (n^2 - 1)^2}{3c^3 N_0} \overline{\mathfrak{E}_z^2},$$

and remembering that

$$I_0 = \frac{c}{4\pi} \overline{\mathfrak{E}_z^2}$$

we obtain equation (7), the application of which to molecular gases is thereby justified.

22. Kerr and Faraday Effect.

It is a well-known fact that external fields are capable of influencing the propagation of light in a gas. In the case of an electric field the resulting phenomenon is called a *Kerr effect*, in the case of a magnetic field a *Faraday effect*.

The general *Kerr effect* can be reduced to the behaviour of monochromatic light traversing the gas at right angles to the direction of the electric field and plane polarised either parallel or perpendicular to it. Then it is found that the index of refraction, or, what comes to the same thing, the phase velocity, is different for the two directions of polarisation. In consequence light originally plane polarised under an oblique angle with respect to the field will in general be elliptically polarised after traversing the gas, one of the two plane polarised components having got ahead in phase of the other. Calling the index of refraction for the parallel component n_π, that for the perpendicular component n_σ, experiment [see e.g. Szivessy (301)] shows that the difference $(n_\pi - n_\sigma)$ has the following properties:

1. It is proportional to $|\mathfrak{E}|^2$, i.e. the square of the strength of the external electric field.

2. It is proportional to N_0, the number of molecules per unit volume of the gas.

3. It generally decreases with increasing temperature T.

4. In the optical region of the spectrum it increases with increasing frequency ν.

5. It is much larger for molecules having a permanent electric moment than for those which have not.

For the description of the experimental material it is customary to introduce the quantity

$$D = \frac{n_\pi - n_\sigma}{\lambda \, |\mathfrak{E}|^2},$$

known as the *Kerr constant* of the gas, where λ denotes the wavelength. According to 1 it is independent of the strength of the electric field.

The ideas which form the basis of the results enumerated above had already been summarised very lucidly by Herzfeld (290) before the advent of the modern quantum theory. We have seen in Art. 6 that under the influence of an electric field a state of a molecule is in general split into a number of component levels distinguished by a quantum number M, which measures the component of angular momentum of the molecule in the direction of the field. We have also seen in Art. 20 that when a molecule in the state q, J, M is subjected to radiation plane polarised in the direction of the field, only those transitions contribute to the scattering moment of the molecule which connect the level q, J, M with a component level having the same M of another state q', J'. By quite similar reasoning we could also have shown that if the light is polarised at right angles to the field, only transitions in which M changes by ± 1 contribute to the scattering moment. Now, as we have seen in Art. 19, the transitions in which M remains unchanged are those which give rise to the Stark effect components polarised parallel to the field of the line q, $J - q'$, J', while the transitions in which M changes by ± 1 represent the perpendicular components in the Stark effect. We may thus say briefly that for radiation polarised parallel to the field only the parallel Stark effect components contribute to the scattering moment and hence according to Art. 21 to the index of refraction n, while for radiation polarised at right angles to the field only the perpendicular Stark effect components contribute to the index of refraction n.

For a vanishing field the parallel and perpendicular components in the Stark effect coincide. Moreover, the sum of the intensities of all the perpendicular components belonging to a given line q, $J - q'$, J' is then equal to the sum of the intensities of all the parallel components so that the line as a whole is unpolarised. The presence of the field affects this situation in two ways. It separates the perpendicular from the parallel components and it modifies their intensities so that the total polarisation is no longer zero. The second change is brought about by the field causing the molecules to accumulate in the component levels M having the smallest energy values and also by its altering the transition probabilities for the various component transitions through the influence which it exerts on the molecular motion.

If the Stark effect is a linear one, it seems at first as if the Kerr effect too should be proportional to the field strength instead of to its square. However, in the linear Stark effect the parallel and perpendicular components are both symmetrically grouped around the position of the originally unresolved line. The "centre of gravity" of the perpendicular components hence coincides still with that of the parallel components as long as only effects linear in $\mathfrak{E}$ are taken into account. Moreover here as well as in the quadratic Stark effect the difference between the intensity of all the perpendicular components and that of all the parallel components is also only of second order in $\mathfrak{E}$, as may be shown by a somewhat cumbersome perturbation calculus. The proportionality of the Kerr effect with $|\mathfrak{E}|^2$ can thus be understood.

The proportionality of the quantity $(n_\pi - n_\sigma)$ with N_0, the number of molecules per unit volume, is immediately evident since n_π and n_σ are separately proportional to N_0. That it in general decreases with increasing temperature T is intimately connected with the statement under 5. We have seen in Art. 6 that appreciable Stark effects occur only in molecules with a permanent electric moment, since in these it is much easier for the electric field to affect the molecular motion. 5 finds thereby its explanation. But we also see from equation (1), Art. 6, that this Stark effect gets smaller the larger the rotational quantum number J of the molecule under consideration. For increasing temperatures more and more molecules are transferred to high rotational states and the resulting decrease of the average Stark effect has as a consequence a corresponding decrease of the Kerr effect.

The increase of the Kerr effect with increasing frequency if ν lies in the optical region results simply from the fact that the important transitions $\nu_{jj'}$, which contribute most to n_π and n_σ, lie for all gases in the ultra-violet and that on account of the resonance denominator $\nu^2_{jj'} - \nu^2$ in equation (4), Art. 20, both n_π and n_σ, and with them their difference, increase on ν approaching $\nu_{jj'}$.

An evaluation of D as a function of T and ν in individual cases, although possible in principle, seems hardly practicable since too little is known about the Stark effect of the different transitions, particularly of those belonging to the continuous part of the spectrum.

We come now to a discussion of the *Faraday effect*. This can be reduced to the behaviour of circularly polarised light propagated in the direction of the applied magnetic field. Here it is found that the index of refraction for right circular light n_ρ is different from that for left circular light n_λ. For that reason if plane polarised light traverses the gas in the direction of the field, there results a rotation of the plane of polarisation. Experiment [for a summary of the empirical data for gases see Darwin and Watson (285)] shows that $(n_\rho - n_\lambda)$ has the following properties:

1. It is proportional to $|\mathfrak{H}|$, the magnitude of the applied magnetic field.

2. It is proportional to N_0.

3. It varies in general little with temperature T.

4. In the optical region of the spectrum it increases with increasing frequency ν.

Again one introduces for convenience a constant

$$G = \frac{(n_\rho - n_\lambda)\,\pi}{\lambda\,|\mathfrak{H}|},$$

Verdet's constant, which according to 1 is independent of the strength of the magnetic field.

The interpretation of the Faraday effect and its characteristic properties follows along the same lines as that given for the Kerr effect. Here it is found that for right circular light propagated parallel to the field only those transitions for which M increases by unity contribute essentially to the scattering moment and hence to the index of refraction n_ρ, and that are those which according to Art. 19 give rise to the right circular components of the Zeeman effect. Similarly if left circular light is used, the left circular components of the Zeeman effect are the ones which contribute to n_λ.

Since in the Zeeman effect the splitting caused by the magnetic field in many of the lines is linear in $|\mathfrak{H}|$, and since here the right circular components and the left circular components are grouped in such a way around the originally unresolved line that to every right circular component on one side there corresponds a symmetrically situated left circular component on the other side, there will in general result also a linear Faraday effect.

The proportionality of G with N_0 and its dependence on ν follow by arguments analogous to those used in the case of an electric field. The independence of the temperature is a consequence of the fact that most gases have no magnetic moment in their normal state, so that the redistribution of the molecules over the various rotational levels produced by a change in temperature has little effect on the influence which the magnetic field exerts on the molecular motion. Again the evaluation of G as a function of T and ν seems at present hardly feasible.

23. DIELECTRIC CONSTANTS.

Under the influence of a constant electric field gases become electrically polarised. The *polarisation*, besides depending on the nature of the gas, is a function of its pressure and temperature as well as of the strength of the electric field, and it is the object of this article to study how it varies with these factors under the conditions which are usually prevalent in experiment. Most of the subsequent discussion applies to all kinds of gases, monatomic as well as molecular, and the word molecule is to be taken in the general sense of atom, diatomic or polyatomic molecule unless specifically stated otherwise.

At ordinary temperatures the molecules of a gas will in general not all be in the same stationary state. For according to Boltzmann's distribution law, equation (2), Art. 21, molecules will be present in appreciable numbers in all those stationary states the energy of which does not differ from that of the lowest state by amounts of a greater order of magnitude than kT, k being Boltzmann's constant, given by equation (3), Art. 21, and T the absolute temperature; and in general for molecular gases the energy differences between successive rotational levels are much smaller than kT if T is of the order of 300° abs.

The problem of finding the electric polarisation per unit volume of the gas is according to the above remarks reduced to determining the constant component of the electric moment parallel to the external field induced in a single molecule for all the stationary states which are realised in the gas at ordinary temperatures. If we denote such a stationary state by j, and if we assume the electric field to be parallel to the direction z, then in the notation introduced in equation (5), Art. 14, the z-component of that part of the electric moment which is independent of the time should be called $\mathfrak{P}_z(j, j')$. This quantity is related to the energy of the molecule in the state j, W_j, which depends on the strength $|\mathfrak{E}|$ of the electric field, by the equation

$$\mathfrak{P}_z(j; j) = -\frac{\partial W_j}{\partial |\mathfrak{E}|}, \quad \dots\dots\dots\dots\dots(1)$$

since

$$W_j = -\int \mathfrak{P}_z(j; j)\, d\,|\mathfrak{E}|,$$

so that in order to find the electric moment of the molecule in the state j parallel to the field it is sufficient to know the energy W_j as a function of $|\mathfrak{E}|$.

Since in general the forces exerted by the external field on the charged particles of which the molecule is composed are small compared to the forces between these particles, it proves convenient to suppose the energy of the state j expanded in powers of $|\mathfrak{E}|$, thus

$$W_j = W_j{}^0 + W_j{}^1 |\mathfrak{E}| + W_j{}^{(2)} |\mathfrak{E}|^2 + \dots, \qquad \dots\dots(2)$$

and to determine the coefficients $W_j{}^1$, $W_j{}^{(2)}$, ... by a perturbation calculus if $W_j{}^0$, the energy of the molecule in the absence of the external field, is known.

The perturbation theory teaches that $W_j{}^1$ and $W_j{}^{(2)}$ are given by

$$W_j{}^1 = -\mathfrak{P}_z{}^0(j;\, j), \qquad \dots\dots\dots(3)$$

$$W_j{}^{(2)} = \sum_{j'}{}' \frac{\mathfrak{P}_z{}^0(j;\, j')\, \mathfrak{P}_z{}^0(j';\, j)}{W_j{}^0 - W_{j'}{}^0}, \qquad \dots\dots(4)$$

where the quantities $\mathfrak{P}_z{}^0(j;\, j')$ are the matrix components of the dipole moment of the molecule in the absence of the external field, determined by equation (5), Art. 14, if we substitute for Ψ_j and $\Psi_{j'}$ the wave functions corresponding to zero field. The prime at the summation sign in equation (4) indicates that the value $j' = j$ is to be omitted in the summation. The quantities $W_j{}^{(3)}$, $W_j{}^{(4)}$, ... are also expressible in terms of $W_j{}^0$ and $\mathfrak{P}_z{}^0(j;\, j')$ by formulae of increasing complication. Since it usually is sufficient to consider $W_j{}^1$ and $W_j{}^{(2)}$, we omit here the expressions for $W_j{}^{(3)}$, etc.

Considering only the quantities $W_j{}^1$ and $W_j{}^{(2)}$ means, according to equation (1), that in the expression for the electric moment of a molecule parallel to the field we confine our attention to the constant terms and those linear in $|\mathfrak{E}|$. We inquire first into that part of the total polarisation per unit volume of the gas which results from these terms. This problem has been treated by Van Vleck [321], [322] under very general assumptions regarding the nature of the molecules composing the gas, and we proceed now to outline briefly his method, referring the reader for the details of the derivation to the original papers.

Let us take any stationary state of the molecule. If we imagine the nuclear structure arrested in its rotation, it may happen that the resulting system has in its electric moment a term independent of the time, a permanent moment, which, just as in Art. 6, we denote by $|\mathfrak{P}|$. In Van Vleck's derivation it is presupposed in the first place that this permanent moment has the same value for all stationary states realised to an appreciable extent in the gas at the temperature T. This condition will be fulfilled with great approximation for all gases at ordinary temperatures. For monatomic gases it is evidently satisfied because atoms have a zero moment in all their stationary states. In molecular gases on the other hand the stationary states present at ordinary temperatures will differ only by their rotational quantum number, by the quantum numbers specifying the orientation of the spin with respect to the molecule, and perhaps, at very high temperatures, by their vibrational quantum numbers, while the electronic quantum numbers are always those of the normal electronic state, the excited ones lying much too high to occur. But the rotation, the spin, and even the vibration do not affect the value of the permanent moment appreciably, which is a quantity depending essentially on the electronic motion.

Between some of the levels which are present at the temperature T radiative transitions may be possible in the absence of the field, or, what comes to the same thing according to equation (1), Art. 14, for certain pairs j and j' among these levels the quantities $\mathfrak{P}_z{}^0(j;j')$ do not vanish. The second assumption made in Van Vleck's paper is that the energy difference between any two levels j and j', realised in the gas at the temperature T and having $\mathfrak{P}_z{}^0(j;j')$ different from zero, is small compared to kT. This condition will be approximately fulfilled for all gases at temperatures not lower than about $200°$ abs. and not high enough to excite the nuclear vibrations, i.e. at temperatures such as are usually prevalent in determinations of the electrical polarisation. For them the energy difference of levels actually present which combine is at most that of a pair having different rotational quantum numbers and, on account of the selection rule for the rotational quantum number, being neighbouring rotational levels. But for these the above condition is in general quite well satisfied at temperatures above 200° abs., the better the larger the moment of inertia of the molecule.

The levels actually present at the temperature T, besides combining among themselves, will be able to combine with other levels not realised in the gas at that temperature. In order that the derivation of Van Vleck may apply, it is necessary that the energy differences of such pairs of combining levels be large compared to kT. At ordinary temperatures these higher levels are those which differ in their vibrational quantum number or in their electronic quantum numbers or in both from the normal state. But at temperatures in the neighbourhood of 300° abs. the corresponding energy differences are usually quite large compared to kT, so that the above requirement is satisfied.

On the basis of the three assumptions just described Van Vleck (321), (322) was able to show that the polarisation produced by unit field per unit volume of the gas is given by

$$\frac{\epsilon - 1}{4\pi} = N_0 \left(\alpha + \frac{|\mathfrak{P}|^2}{3kT} \right), \quad \dots\dots\dots\dots(5)$$

where N_0 denotes the number of molecules per unit volume of the gas and α a constant independent of the temperature. The quantity ϵ is the *dielectric constant* of the gas, since the ratio of the polarisation per unit volume to the field strength is equal to $(\epsilon - 1)/4\pi$, provided the density of the substance is sufficiently small so that the average influence which the molecules exert on each other due to their polarisation is small compared to that of the external field, a condition satisfied in gases.

Equation (5) is the same as that originally derived by Debye (308) on the classical theory for molecules with a permanent electric moment. We see from it that the polarisation is proportional to the density of the gas and that it consists of two parts, one independent of the temperature, the other inversely proportional to it. The latter, in particular, vanishes if the molecule has no permanent electric moment.

Before the general proof of equation (5) was given by Van Vleck, the behaviour of a number of special models had been considered on the basis of quantum mechanics. Thus Mensing and Pauli (316), Van Vleck (320), Kronig (311), and Manneback (314) had shown that for a gas composed of rotators carrying a dipole moment there results, for the polarisation at temperatures not too low, the second term in equation (5). The same result was obtained by Kronig (312), Debye and Manneback (309), and Manneback (315) for a gas composed of symmetrical tops carrying a permanent electric moment parallel to their axis of symmetry.

At low temperatures the second assumption made in deriving equation (5) will no longer be satisfied with great accuracy. Then the polarisation per unit volume per unit field, instead of being given by equation (5), will be determined by

$$\frac{\epsilon - 1}{4\pi} = N_0 \alpha + \frac{N_0 |\mathfrak{P}|^2}{3kT} [1 - f(T)], \quad \dots\dots\dots(6)$$

where $f(T)$ is a series in powers of $1/T$ and hence vanishing for large values of T. The first term in $f(T)$ has been computed for a gas composed of rotators with a dipole moment by Kronig (311), and for a gas composed of symmetrical tops with a dipole moment along the axis of symmetry by Manneback (315). Later Van Vleck (322) has given an expression for the first term in $f(T)$ for gases composed of asymmetrical tops. He finds for it

$$\frac{h^2}{48\pi^2 kT |\mathfrak{P}|^2} \left[\mathfrak{P}_\xi^2 \left(\frac{1}{A_\eta} + \frac{1}{A_\zeta} \right) + \mathfrak{P}_\eta^2 \left(\frac{1}{A_\zeta} + \frac{1}{A_\xi} \right) + \mathfrak{P}_\zeta^2 \left(\frac{1}{A_\xi} + \frac{1}{A_\eta} \right) \right],$$

where $\mathfrak{P}_\xi$, $\mathfrak{P}_\eta$, $\mathfrak{P}_\zeta$ are the components of the permanent moment along the principal axes of inertia ξ, η, ζ of the molecule and A_ξ, A_η, A_ζ the principal moments of inertia. The results of Kronig and of Manneback are obtained from this general expression by putting $\mathfrak{P}_\xi = \mathfrak{P}_\eta = 0$, $\mathfrak{P}_\zeta = |\mathfrak{P}|$, and $A_\xi = A_\eta$.

Hitherto we have considered in the development (2) of W_j in powers of $|\mathfrak{E}|$ only terms containing no power higher than the second. For very strong fields the higher terms will become of

importance; *saturation* will set in. Niessen (317) has investigated
the behaviour of rigid molecules carrying a permanent moment for
fields of any strength. He showed that if the permanent moment
is the same for all states of the molecule realised at the temperature
T to an appreciable extent and if the energy differences between
combining states are small compared to kT, then the polarisation
per unit volume per unit field is given by the generalised formula
of Debye

$$\frac{\epsilon - 1}{4\pi} = N_0 \,|\,\mathfrak{P}\,| \left(\coth \frac{|\,\mathfrak{P}\,|\,|\,\mathfrak{E}\,|}{kT} - \frac{kT}{|\,\mathfrak{P}\,|\,|\,\mathfrak{E}\,|} \right), \quad ......(7)$$

which for small values of $|\,\mathfrak{E}\,|$ reduces to equation (5). He also
investigated the polarisation of a gas composed of deformable
molecules up to terms in $|\,\mathfrak{E}\,|^3$ and obtained

$$\frac{\epsilon - 1}{4\pi} = N_0 \left(\alpha + \frac{|\,\mathfrak{P}\,|^2}{3kT} \right) + \left(\alpha' + \frac{\alpha''}{kT} + \frac{\alpha'''}{k^2 T^2} - \frac{|\,\mathfrak{P}\,|^4}{45 k^3 T^3} \right) |\,\mathfrak{E}\,|^2, \,...(8)$$

where the α's are constants independent of T, determined by the
nature of the molecule.

It is interesting to note a difference here between the classical
and the quantum theory. In the classical theory equation (7) holds
for all values of $|\,\mathfrak{E}\,|$ and T. Saturation may be produced either by
making $|\,\mathfrak{E}\,|$ large or T small, since in equation (7) only the ratio
of $|\,\mathfrak{E}\,|/T$ enters. In the quantum theory equation (7) no longer
holds at low temperatures, and indeed it is found in the special
case of a gas composed of rotators carrying a dipole moment that

$$\frac{\epsilon - 1}{4\pi} = N_0 \frac{8\pi^2 A \,|\,\mathfrak{P}\,|^2}{3h^2}$$

at absolute zero, A being the moment of inertia of the rotators,
while in the classical theory this quantity becomes infinite.

The experimental material confirms in general equation (5).
Temperatures are usually not low enough and the field strength
not high enough to require the corrections given in (6), (7) and (8)
to be taken into account. The experimentally established fact that
diatomic gases with homonuclear molecules have a vanishing

permanent moment follows theoretically from the symmetry properties of the wave functions against interchange of the nuclei discussed in Art. 9. For a bibliography of the experimental evidence on dipole moments the reader is referred to a comprehensive survey by Blüh (307) and the dissertation of Højendahl (310).

It may be mentioned in this connection that the expressions for the energy of a diatomic molecule in an electric field given in Art. 6 when discussing the linear Stark effect are those furnished by equation (3). The values derived from them by equation (1) for the constant component of the electric moment of the molecule in the direction of the field may be tested directly by sending a narrow beam composed of the molecules through a non-uniform electric field and measuring the deflection produced. Experiments of this kind have been performed by Wrede (328).

24. Magnetic Susceptibilities.

Just as an electric field polarises a gas electrically, a magnetic field will cause it to become magnetised. Again the *magnetisation* for a given gas is a function of its pressure and temperature and the strength of the applied field, and we wish in this article to investigate its dependence on these variables, confining our attention to molecular gases.

Instead of having to find the electric moment we must now determine the constant component of the magnetic moment induced in a single molecule by the magnetic field parallel to its direction for all those stationary states j which contain molecules in appreciable numbers at the temperature T. Denoting the magnetic moment, which in quantum mechanics just like the electric moment is a vector matrix, by $\mathfrak{M}$, we must obtain its component $\mathfrak{M}_z(j; j)$ if the external field is in the z-direction. In analogy with equation (1), Art. 23, it is given by

$$\mathfrak{M}_z(j; j) = -\frac{\partial W_j}{\partial |\mathfrak{H}|},$$

where W_j is the energy of the state j, depending of course on the strength $|\mathfrak{H}|$ of the magnetic field.

If now we suppose W_j expanded in powers of $|\mathfrak{H}|$, thus

$$W_j = W_j{}^0 + W_j{}^1 |\mathfrak{H}| + W_j{}^{(2)} |\mathfrak{H}|^2 + \dots,$$

then $W_j{}^1$, $W_j{}^{(2)}$, ... are again given by the perturbation theory. In particular, as shown e.g. by Van Vleck (324),

$$W_j{}^1 = - \mathfrak{R}_z{}^0(j; j),$$

$$W_j{}^{(2)} = \sum_{j'}{}' \frac{\mathfrak{R}_z{}^0(j; j')\, \mathfrak{R}_z{}^0(j'; j)}{W_j{}^0 - W_{j'}{}^0} + \frac{e^2}{8mc^2} \sum_r \rho_r{}^2(j; j). \quad \dots(1)$$

Here $\mathfrak{R}_z{}^0(j; j')$ is the matrix component of the magnetic moment of the molecule in the absence of the magnetic field, defined by an equation analogous to equation (5), Art. 14, for the electric moment with wave functions referring to a vanishing field. In the summation of the first term in equation (1) the value $j' = j$ is to be omitted as indicated by the prime at the summation sign. ρ_r is the distance of the r-th electron from an axis drawn through the centre of mass of the molecule parallel to the direction of the field, and $\rho_r{}^2(j; j')$ is the matrix component of its square belonging to the states j and j'. To be exact there arises a contribution to the second term in equation (1) from the nuclei, which however on account of their large mass is negligible. Finally $W_j{}^{(3)}$ etc. are also expressible in terms of the same quantities as $W_j{}^1$ and $W_j{}^{(2)}$.

The problem of the magnetisation of gases too has been treated by Van Vleck (321), (324) by the same general method as that of the electric polarisation. By letting $\mathfrak{R}$ take the place of $\mathfrak{P}$ and by making assumptions regarding its permanency and the energy differences between the stationary states analogous to those of Art. 23, he showed that the magnetisation per unit volume per unit field, or in other words the *magnetic susceptibility*, is given by

$$\kappa = N_0 \left(\beta + \frac{|\mathfrak{R}|^2}{3kT} \right). \quad \dots\dots\dots\dots(2)$$

Here $|\mathfrak{R}|$ is the permanent magnetic moment and β a constant independent of temperature, which in general is negative. According to equation (2) the susceptibility is proportional to the pressure and consists of two parts, one independent of temperature, the other changing inversely as the temperature and being present only if there is a permanent magnetic moment. Equation (2) is identical with the formula first derived by Langevin (313) on the basis of the classical theory.

Most molecular gases have no permanent magnetic moment in the stationary states present at ordinary temperatures. Since, as mentioned before, β in equation (2) may generally be expected to be negative, κ will be then also negative; the gases are *diamagnetic*. To determine the value of κ exactly from theoretical considerations will not be possible on account of the mathematical difficulties except in the simplest cases. For H_2 an approximate calculation has been given by Wang (326) and by Van Vleck and Miss Frank (325). Among the diatomic gases O_2 and NO are the only ones which have a permanent magnetic moment in the states present at ordinary temperatures, among the polyatomic gases ClO_2 is one of the few. These gases will be *paramagnetic*.

O_2 has as its normal state a $^3\Sigma$-state, with zero orbital angular momentum of the electrons around the internuclear line and a spin $S = 1$. The square of the magnetic moment of the spin has according to wave mechanics the magnitude $4S(S+1)$, using the Bohr magneton as the unit to measure magnetic moment. If this is substituted in equation (2), there results for the volume susceptibility of O_2 at 20° C. and 76 cm. pressure, neglecting the term with β, the value $0·142 . 10^{-6}$, while the experimental value is $0·1434 . 10^{-6}$ according to Bauer and Piccard (305) and $0·1447 . 10^{-6}$ according to Wills and Hector (327). For ClO_2 the assumption $S = \frac{1}{2}$ gives for the volume susceptibility at 20° C. and 76 cm. pressure the value $0·0534 . 10^{-6}$ if again the term with β is neglected in equation (2), while Taylor (319) finds experimentally the value $0·0557 . 10^{-6}$, the discrepancy lying probably within the limits of error of Taylor's measurements.

The susceptibility of NO requires special consideration. For according to the spectroscopic data the normal state of NO is a $^2\Pi$-state, consisting of two levels $^2\Pi_{\frac{1}{2}}$ and $^2\Pi_{\frac{3}{2}}$, the former being the one of smaller energy. For $^2\Pi_{\frac{1}{2}}$ we have a component of orbital angular momentum of the electrons along the internuclear line of one unit and an oppositely directed spin component of half a unit. Since the ratio of magnetic moment to angular momentum is twice as great for the spin as for the orbital angular momentum, the resultant magnetic moment parallel to the internuclear line vanishes for the state $^2\Pi_{\frac{1}{2}}$. For the state $^2\Pi_{\frac{3}{2}}$ the two angular momenta are added instead of being subtracted, and since with one unit of orbital angular momentum there is associated a magnetic moment of one Bohr magneton, the total magnetic moment parallel to the internuclear line is here two Bohr magnetons. The separation of the component levels $^2\Pi_{\frac{1}{2}}$ and $^2\Pi_{\frac{3}{2}}$ is of the same order of magnitude as kT at ordinary temperatures, and both are hence realised in the gas. For NO the condition of permanency of the magnetic moment is thus not fulfilled, the various states actually present not all having the same magnetic moment. Furthermore, the matrix element of the magnetic moment $\Re(j; j')$ belonging to the two component levels does not vanish, and since the separation of these levels at ordinary temperatures is not small compared to kT, the requirement regarding the energy difference between "combining" levels among those actually present is not satisfied.

Equation (2) can therefore not be applied to NO so that special considerations become necessary for this gas which have also been given by Van Vleck (324). He finds for the susceptibility

$$\kappa = \frac{N_0 \delta^2 |\Re_0|^2}{3kT},$$

where δ now is not a constant independent of T, but given by

$$\delta^2 = 4\frac{1 - e^{-x} + xe^{-x}}{x + xe^{-x}}$$

and $|\mathfrak{R}_0|$ the Bohr magneton. x is an abbreviation for the ratio of the doublet interval of the two component levels $^2\Pi_{\frac{1}{2}}$ and $^2\Pi_{\frac{3}{2}}$ to kT:

$$x = \frac{h\Delta\nu}{kT} = \frac{173\cdot2}{T}.$$

The table below, taken from Van Vleck's paper, shows the variation of δ with T.

T	0	50	100	175	250	293	350	500	1000	∞
δ	0	1·098	1·489	1·713	1·806	1·836	1·864	1·908	1·954	2·000

Thus far the variation of the susceptibility with T has not been studied and there are only a few isolated values available which agree well with the theory. Thus Bauer and Piccard (305) and Soné (318) find at 20° C. and 76 cm. pressure $\kappa = 0\cdot061 \cdot 10^{-6}$, while the theoretical value is $0\cdot060 \cdot 10^{-6}$. Measurements of Bitter (306) also confirm the theory.

For the case of diatomic gases Niessen (317) has studied the phenomena to be expected when the strength of the magnetic field becomes very large so that *saturation* sets in. He confines his attention to the case where the multiplet intervals, if there are any, are either small or large compared to kT. O_2 is the only example for the first case, while no example is known for the second case, the doublet interval of NO being comparable to kT at ordinary temperatures. For O_2 the susceptibility is given by

$$\kappa = N_0\left[3\,|\mathfrak{R}_0|\left(\coth\frac{3\,|\mathfrak{R}_0||\mathfrak{H}|}{kT} - \frac{kT}{3\,|\mathfrak{R}_0||\mathfrak{H}|}\right)\right.$$
$$\left. - |\mathfrak{R}_0|\left(\coth\frac{|\mathfrak{R}_0||\mathfrak{H}|}{kT} - \frac{kT}{|\mathfrak{R}_0||\mathfrak{H}|}\right)\right].$$

25. SPECIFIC HEATS.

The *specific heat* per gram molecule of a gas at constant volume is given by

$$C_v = \left(\frac{\partial E}{\partial T}\right)_v, \qquad \dots\dots\dots\dots\dots(1)$$

E being the energy of one gram molecule of the gas, T the absolute temperature, while the differentiation is to be performed at constant volume.

Throughout this section we assume that the interaction between the molecules of the gas is negligible. This does not only mean that the forces between the molecules play a subordinate rôle, but also that the *degeneration phenomena* investigated by Einstein are of no importance in the region of temperature and pressure to which our discussion refers. In practice for all diatomic gases these phenomena would set in at temperatures so low or pressures so high that the intermolecular forces would be appreciable too, so that when the latter may be disregarded, degeneration will also be absent. Under this condition C_v is the sum of the translational energy of the molecules and their internal energy. The *translational energy* per gram molecule is given by $\frac{3}{2}RT$, where

$$R = Nk = 8\overset{!}{\cdot}314 \, . \, 10^7 \text{ erg/degree} \quad \dots\dots\dots\dots(2)$$

is the gas constant per gram molecule,

$$N = 6\cdot064 \, . \, 10^{23}$$

being Avogadro's number, i.e. the number of molecules per gram molecule, k Boltzmann's constant, given by equation (3), Art. 21. Denoting by N_j the number of molecules per gram molecule in the stationary state j with energy W_j, the *internal energy* per gram molecule evidently is $\sum_j N_j W_j$ so that

$$E = \frac{3}{2}RT + \sum_j N_j W_j. \quad \dots\dots\dots\dots\dots(3)$$

Now according to Boltzmann's distribution law

$$N_j = \frac{N}{Z} g_j e^{-\frac{W_j}{kT}}, \quad \dots\dots\dots\dots\dots\dots(4)$$

where g_j is the statistical weight of the state j and

$$Z = \sum_j g_j e^{-\frac{W_j}{kT}}. \quad \dots\dots\dots\dots\dots\dots(5)$$

From equations (2) to (5) it follows that

$$E = \tfrac{3}{2}RT + RT^2 \frac{d}{dT}\log Z.$$

Equation (1) gives then

$$C_v = \tfrac{3}{2}R + R\frac{d}{dT}\left(T^2\frac{d}{dT}\log Z\right) \quad \dots\dots\dots\dots(6)$$

so that C_v may be computed if Z is known as a function of T.

Only those stationary states of the molecules in a molecular gas contribute to Z which are realised in appreciable numbers at the temperature T. At temperatures available in the laboratory excited electronic states of the molecules need thus not be considered since they lie much too high. We discuss now in order the various possible cases.

1. *Temperatures low enough so that the nuclear vibrations are not excited.*

(a) *Heteronuclear diatomic molecules in a $^1\Sigma$-state.* All hetero-nuclear diatomic molecules for which specific heat measurements appear feasible with the exception of NO belong to this category. To Z there contribute the various rotational levels J of the normal electronic state with a statistical weight $2J + 1$ and, according to Art. 5, an approximate energy

$$W_J = B_0\,(J + \tfrac{1}{2})^2.$$

We have thus

$$Z = \sum_{J=0}^{\infty} (2J + 1)\,e^{-\frac{B_0}{kT}(J+\frac{1}{2})^2}.$$

If B_0 is small compared to kT, which is the case in all measurements for this class of molecular gases, Z according to Mulholland (335) can be written as the series

$$Z = \frac{kT}{B_0} + \frac{1}{12} + \frac{7}{480}\frac{B_0}{kT} + \dots.$$

Under the experimental conditions only the first term needs to be considered, and substitution in equation (6) gives

$$C_v = \tfrac{5}{2}R$$

in good agreement with the empirical data (see below).

(b) *Homonuclear diatomic molecules in a* $^1\Sigma$-*state.* All homo-nuclear diatomic molecules for which the specific heat has been measured with the exception of O_2 belong to this group. The only difference between these molecules and those discussed under (a) is that the rotational levels with even J and those with odd J have now statistical weights $g_e(2J+1)$ and $g_o(2J+1)$ respectively as explained in Art. 18, the values of g_e and g_o depending on the magnitude of the nuclear spin. We have thus

$$Z = \sum_{\text{even } J} g_e(2J+1) e^{-\frac{B_0}{kT}(J+\frac{1}{2})^2} + \sum_{\text{odd } J} g_o(2J+1) e^{-\frac{B_0}{kT}(J+\frac{1}{2})^2}. \quad (7)$$

If B_0 is small compared to kT, which is the case in all experimental determinations of the specific heat except for H_2 at low temperatures, then the procedure of Mulholland gives for Z in first approximation

$$Z = \text{const. } T,$$

which according to equation (6) leads again to the value

$$C_v = \tfrac{5}{2} R$$

in agreement with the empirical results (see below).

The case of H_2 requires special consideration since here measurements are available in a region of T, where B_0 is no longer small compared to kT. We have seen in Art. 18 that for H_2 g_e and g_o are respectively 1 and 3. If, however, one computes from the expression (7) with these values of g_e and g_o the specific heat by means of equation (6), one finds a severe disagreement between the theoretical and experimental values.

An explanation of this discrepancy was given by Dennison (332). He pointed out that transitions between the even and the odd rotational levels will be exceedingly rare, a phenomenon which we discussed in section 2 of Art. 15. At ordinary temperatures where kT is large compared to B_0 so that many rotational levels are excited there will be three times as many H_2-molecules in rotational levels with odd than in those with even J. Now, if the temperature is lowered into the region where kT no longer is large compared to B_0, this ratio should be different under the condition of temperature equilibrium, the levels with even J getting a greater share of the molecules than previously since on the average they lie lower. At absolute zero in fact all molecules should be in the state $J = 0$. However, on account of the small probability of transitions between levels with even and odd J, temperature equilibrium will not be reached during the time that the temperature is lowered and the specific heat measured. (In a paper by Bonhoeffer and Harteck (331)

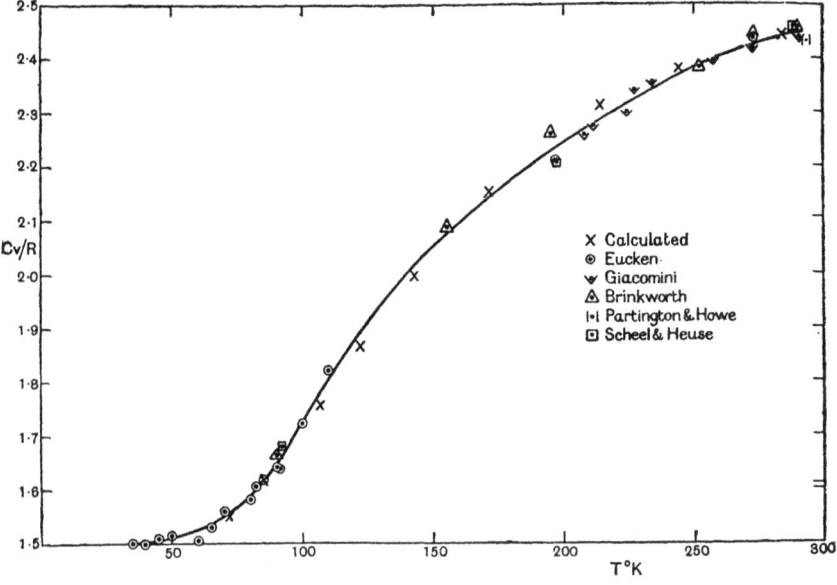

Fig. 14. The specific heat of hydrogen.

calculations of Wigner are given, according to which the establishment of thermodynamic equilibrium should take years under ordinary conditions.) We may assume with safety that the ratio actually remains $1 : 3$. The result will be that H_2 behaves as a mixture of two component gases in the proportion $1 : 3$, the one having only rotational levels with even J, the other only those with odd J and each by itself being in thermal equilibrium. The results obtained for the specific heat on this basis agree well with the experimental determination as shown by Fig. 14.

There naturally arises the question if it would not be possible to accelerate the transformation of one component of H_2 into the other. If that could be done, a change in the specific heat would be expected. This conclusion was indeed confirmed by measurements of Bonhoeffer and Harteck (329), (330), (331) and of Eucken (333). By condensing H_2 on charcoal and then evaporating it again, in the absence of the charcoal these authors have obtained values for the thermal conductivity and the specific heat of the gas differing from those found if the gas had simply been cooled to the same temperature without condensation. The charcoal acted as a catalyst and helped to establish thermal equilibrium between the two components. At very low temperatures all H_2-molecules can thus be brought into the state $J = 0$. If the H_2 is then raised to higher temperatures in the absence of the catalyst, it will consist entirely of molecules with even J. As shown by Bonhoeffer and Harteck (331) this manifests itself in the band spectra, the alternate lines which originally were strong becoming much weaker or disappearing altogether.

(c) *NO*. Here we have at ordinary temperatures the rotational levels belonging to the two components $^2\Pi_{\frac{1}{2}}$ and $^2\Pi_{\frac{3}{2}}$ of the normal $^2\Pi$-state of that molecule, the energy difference of which we shall call ΔW, $^2\Pi_{\frac{1}{2}}$ being the lower component. For $^2\Pi_{\frac{1}{2}}$, J takes the values $\frac{1}{2}, \frac{3}{2}, \ldots$ for $^2\Pi_{\frac{3}{2}}$, J takes the values $\frac{3}{2}, \frac{5}{2}, \ldots$. We get thus for

$$Z = \sum_{J=\frac{1}{2}, \frac{3}{2}, \ldots} (2J+1)e^{-\frac{B_0}{kT}(J+\frac{1}{2})^2} + \sum_{J=\frac{3}{2}, \frac{5}{2}, \ldots} (2J+1)e^{-\frac{1}{kT}[\Delta W + B_0(J+\frac{1}{2})^2]}.$$

The method of Mulholland gives here in first approximation

$$Z = \frac{kT}{B_0}(1 + e^{-\frac{\Delta W}{kT}}).$$

From this one finds for C_v with the help of equation (6)

$$C_v = \tfrac{5}{2}R + R\left(\frac{\Delta W}{kT}\right)^2 \frac{e^{-\frac{\Delta W}{kT}}}{(1 + e^{-\frac{\Delta W}{kT}})^2}.$$

The value of $\Delta W/kT$ has been given in Art. 24. At ordinary temperatures the second term in C_v will lie near the limit of experimental error.

(d) O_2. This gas has as its lowest state a $^3\Sigma$-state. The presence of the spin gives rise to a fine structure of the rotational levels, but this is so small that it will not affect the value of the specific heat, which here as under (b) will be

$$C_v = \tfrac{5}{2}R$$

in the temperature region in which measurements are available. The following table shows the values of the ratio C_p/C_v, the specific heats at constant pressure and at constant volume, which, since from thermodynamics $C_p = C_v + R$, should be equal to $\tfrac{7}{5}$, for various diatomic gases at temperatures at which the vibrations are not excited. In all cases a satisfactory agreement is found.

Observed values of C_p/C_v for diatomic gases.

Gas	$T°$ C.	C_p/C_v	Authority
H_2	16	1·407	
N_2	20	1·398	
	− 181	1·419	
O_2	20	1·398	
	− 76	1·411	
	− 181	1·404	Scheel and Heuse (336)
CO	18	1·396	
	− 180	1·417	
NO	15	1·38	
	− 45	1·39	
	− 80	1·38	
HCl	15	1·40	Partington (336)

(e) *Polyatomic molecules.* If kT is large compared to the energy differences between neighbouring rotational states, one will get just as in the case of diatomic molecules the classical equipartition value for C. Since polyatomic molecules have three rotational degrees of freedom, one finds thus

$$C_v = 3R.$$

Deviations from this value at low temperatures might be observable if the molecule has a small moment of inertia. H_2O, NH_3, and CH_4 are gases fulfilling this condition, and since the first two condense at rather high temperatures, CH_4 is the only one in which an effect could perhaps be detected. At present no very satisfactory measurements at low temperatures appear to be available.

2. *Temperatures high enough so that the nuclear vibrations are excited.*

(a) *Diatomic molecules.* The energy of vibration and rotation is approximately given according to Art. 5 by an expression of the type

$$W = h\nu_0 (v + \tfrac{1}{2}) + B_0 (J + \tfrac{1}{2})^2. \quad \dots\dots\dots\dots(8)$$

Since we are now in a region of high temperatures, B_0 certainly is small compared to kT so that in any case we need to retain only the first term in Mulholland's development. With the above value of W, Z becomes then according to equation (5)

$$Z = \text{const. } T \cdot \frac{1}{1 - e^{-\frac{h\nu_0}{kT}}}.$$

This gives for C_v

$$C_v = \tfrac{5}{2} R + R \left(\frac{h\nu_0}{kT}\right)^2 \frac{e^{-\frac{h\nu_0}{kT}}}{(1 - e^{-\frac{h\nu_0}{kT}})^2}.$$

For very large values of T we find

$$C_v = \tfrac{7}{2} R.$$

The experimental data on specific heats at higher temperatures are rather meagre.

In some cases it may be necessary to consider in the energy the coupling of vibration and rotation. This may be done either by using instead of equation (8) the more complicated expression (2), Art. 5 for the energy or to take the position of the various vibration-rotation levels directly from experimental data. The second method has been used by McCrea [334] for H_2. The following table shows a comparison of observed and computed values of C_v.

Specific heat C_v of H_2 at high temperatures.

$T°$ abs.	C_v obs.	C_v calc.
600	5·08	4·98
800	5·22	5·04
1000	5·36	5·16
1200	5·50	5·34
1600	5·78	5·72
1800	5·92	5·89
2000	6·06	6·05
2500	6·41	6.37

(b) *Polyatomic molecules.* The difficulty of taking into account the vibration for these molecules arises from our inadequate knowledge of the vibrational levels. The experimental data are also rather meagre. The reader is referred to R. H. Fowler, *Statistical Mechanics,* for a critical discussion of the rather uncertain results for NH_3, CH_4, CO_2, and H_2O.

CHAPTER V

MOLECULE FORMATION AND CHEMICAL BINDING

26. HETEROPOLAR MOLECULES.

The progress made on the basis of the spectroscopic evidence in the interpretation of the properties of molecules naturally leads to the question if it is possible to arrive at an understanding of the enormous number of experimental facts accumulated in the course of time by the chemist and expressed by him in terms of *chemical formulae*, the concepts of *valency* and the *saturation of valencies*. Attacks along two different lines have been made on these problems. The first, due to Hund (342), (343), which preceded the development of quantum mechanics, is semi-empirical and applies to *heteropolar molecules* in which the atoms composing them may be conceived of as being essentially in an ionised state. This method has been applied to some of the simpler molecules of this type, notably H_2O, H_2S, H_2Se, and NH_3, and has led to important conclusions regarding their shape. We shall outline it in this article. The second method, due to Heitler and London (338), (339), (340), (346), (347) is based directly on the wave equation of quantum mechanics and applies above all to *homopolar compounds*. Due to the mathematical difficulties it permits at present only for the simple atoms, like H, He, etc., definite predictions regarding their chemical behaviour to be made. We shall devote the two following articles to a survey of it.

In the theory of Hund a heteropolar molecule is considered as being built up of ions; thus in the molecule H_2O we have an ion $O^=$ with two negative charges and two hydrogen nuclei. The question raised is then what will be the *equilibrium configuration of the ions*, and in order to answer it the forces between the ions must be known as a function of the distance of separation. For the molecules which he considers Hund determines these forces as illustrated for the case of H_2O by the subsequent discussion.

The ion O^- has as many electrons as the ion F^- or neutral Ne. The electrons form hence a rare gas configuration in all three atoms, and the only difference is that the nuclear charge for O^- is two units less, for F^- one unit less than the nuclear charge of Ne. Now the forces between the ion F^- and an H-nucleus can be determined rather accurately as a function of the distance of separation ρ. For we know that there exists a molecule HF, showing that the potential energy of the two particles H^+ and F^- has a minimum for a certain value of ρ. The vibrational bands of HF furnish us with the form of the potential energy curve in the neighbourhood of the equilibrium distance. From the energy of dissociation of HF we find how much the minimum value of the potential energy lies below the energy of an H-nucleus and an F-ion separated from each other. Finally at large distances of separation the forces between the two particles will be the sum of the Coulomb forces due to their charges and the forces arising from the polarisation of the F-ion by the H-nucleus. The mutual potential energy for large values of ρ may hence be written in the form

$$V = -\frac{e^2}{\rho} - \frac{\alpha e^2}{2\rho^4}, \qquad \dots\dots\dots\dots\dots\dots(1)$$

where α is the polarisability of the F-ion, which may be determined from data on the optical refractivity. In Fig. 15 we have represented V as a function of ρ by full-drawn curves in the neighbourhood of the equilibrium position and for large values of ρ as it follows by the method just described. As is apparent from their dotted continuations the two curves join on to each other quite well so that the potential energy may be regarded as a known function of ρ for all values of ρ.

The ion O^-, as stressed above, differs from the ion F^- only by having a negative charge by one unit greater, and consequently a somewhat looser binding of its electrons. For the potential energy between the ion O$^-$ and an H-nucleus we may hence write approximately

$$V_0 = \frac{1}{\omega} V_F\left(\frac{\rho}{\omega}\right) - \frac{e^2}{\rho},$$

where ω is a constant to be suitably determined and actually > 1. For then we have for large values of ρ

$$V_0 = -\frac{2e^2}{\rho} - \frac{\alpha \omega^3 e^2}{2\rho^4}, \quad \dots\dots\dots\dots\dots(2)$$

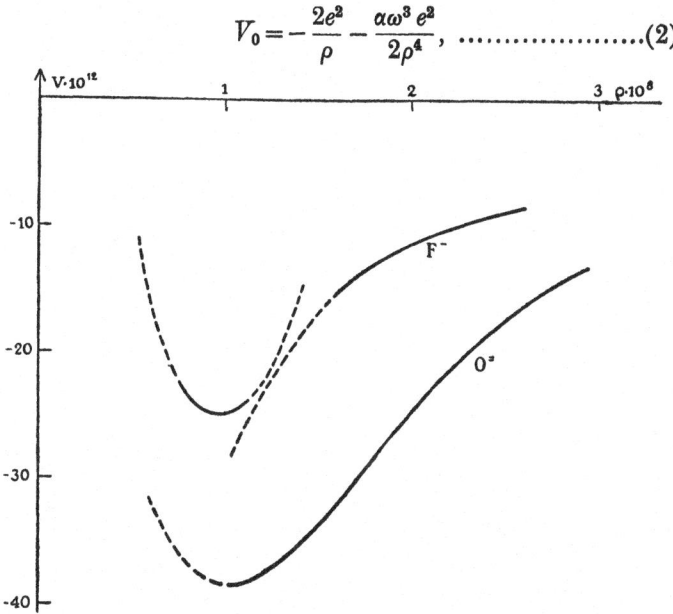

Fig. 15 Potential energy of hydrogen nucleus in the field of fluorine ion
F$^-$ or oxygen ion O$^=$ as function of internuclear distance.

while at smaller distances of separation the potential energy curve will be elongated in the direction ρ with respect to that of F$^-$ corresponding to the fact that the ion O$^-$ is larger in size than the ion F$^-$. From equations (1) and (2) it is apparent that ω^3 must be in the ratio of the polarisabilities of O$^-$ and F$^-$ and can hence be determined from experimental data. We have drawn in Fig. 14 besides the potential energy of H$^+$ with respect to F$^-$ that of H$^+$ with respect to O$^-$, as given by Hund on the basis of the procedure just described.

In considering the $O^=$ ion in the presence of two H-nuclei it must be remembered that the polarisation caused by one of them will be less than the value which would be obtained if the other one were absent, since positive charges strengthen the binding of the electrons. In utilising the results for the mutual potential energy of an $O^=$ ion and an H-nucleus previously obtained for the case of an $O^=$ ion and two H-nuclei a correction becomes necessary. With the corrected curve for the potential energy and a simple electrostatic repulsion between the two H-nuclei Hund investigates the stable equilibrium configuration of the three particles on the basis of classical mechanics.

The most important result of this investigation is that the nuclei in H_2O do not lie in a straight line but form an isosceles triangle with the O-nucleus at the vertex. A similar conclusion results for the molecules H_2S and H_2Se. This expectation is substantiated by the experimental fact that both H_2O and H_2S have permanent electric moments. Hund also calculates the energy necessary for the dissociation of the above molecules.

For NH_3 the same method leads to the result that the nuclei form a pyramid with the H-nuclei at the three corners of an equilateral base and the N-nucleus vertically above the centre of the base. Here too it is possible to compute the energy necessary for dissociation.

27. HOMOPOLAR MOLECULES. CHEMICAL FORCES BETWEEN TWO H-ATOMS AND TWO He-ATOMS.

The simplest case of the chemical interaction between two neutral atoms is the *union of two hydrogen atoms in their normal state* to form a hydrogen molecule. This problem has been investigated by Heitler and London (338). Here we have two equal nuclei and two electrons, one with each nucleus to start with, bound in the normal state, and the question is how the energy of the system changes when we diminish the internuclear distance ρ.

Calling the nuclei I and II and the electrons 1 and 2, we may say that for infinite ρ there are two possibilities: Electron 1 may be with nucleus I and electron 2 with nucleus II or vice versa. Disregarding at first the spin of the electrons, let us denote the solution of the wave equation of the system consisting of nucleus I and electron 1 alone which corresponds to the normal state of the system by Ψ_{I1}. Also let Ψ_{I2} be the solution for the normal state of the system consisting of nucleus I and electron 2, while Ψ_{II1} and Ψ_{II2} have a similar significance for the system built up of nucleus II and the electrons 1 or 2 respectively.

If we consider the four particles together, and if the two nuclei are still very far apart, both $\Psi_{I1} \Psi_{II2}$ and $\Psi_{I2} \Psi_{II1}$ will be solutions of the wave equation for this system as a whole, and they both belong to the same energy value corresponding to one electron being with each nucleus in the state of least energy. This energy value is hence degenerate, any linear combination of $\Psi_{I1} \Psi_{II2}$ and $\Psi_{I2} \Psi_{II1}$ of the form

$$c_1 \Psi_{I1} \Psi_{II2} + c_2 \Psi_{I2} \Psi_{II1}$$

being a solution of the wave equation belonging to it.

When the two nuclei are approached, the degeneracy is removed, the energy level is split into two, and with each component there is associated a particular linear combination of the two solutions The investigation shows that with the one component there is associated the wave function

$$\Psi_{I1} \Psi_{II2} + \Psi_{I2} \Psi_{II1}, \quad \dots\dots\dots\dots\dots(1)$$

which is symmetrical in the coordinates of the electrons so that it is not altered when these are interchanged, while with the other component there is associated the wave function

$$\Psi_{I1} \Psi_{II2} - \Psi_{I2} \Psi_{II1}, \quad \dots\dots\dots\dots\dots(2)$$

antisymmetrical in the positional coordinates of the electrons and hence changing sign when they are interchanged.

The two functions (1) and (2) are exact solutions of the wave equation only for infinitely large values of ρ. When the nuclei are approached, they satisfy equations differing by terms from the actual wave equation which are small as long as the nuclei do not come too close. The influence of these terms may then be taken into account by a perturbation calculation. They modify the wave functions (1) and (2), and they give rise to the splitting of the energy value mentioned above.

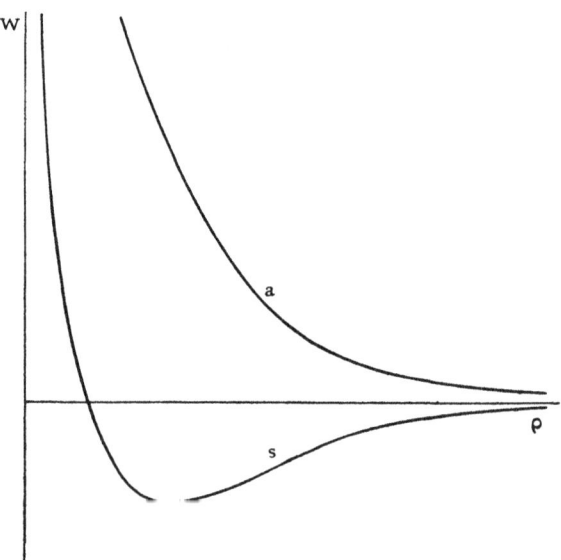

Fig. 16.   Mutual potential energy of two H-atoms for two possible modes of interaction as function of internuclear distance.

If we confine ourselves to first order perturbation effects, it is possible to compute numerically the energy of the two component states as a function of ρ. This has been done by Heitler and London and by Sugiura (349). It is found that the energy of the component level which for large values of ρ is characterised by the symmetrical wave function (1) is represented by the curve s of Fig. 16, while the energy of the component level with the antisymmetrical wave function (2) is represented by the curve a. The former has a minimum so that two hydrogen atoms approaching each other along it can form a stable molecule, while for an approach along the curve a no molecule formation is possible.

We have mentioned repeatedly (see e.g. Art. 3) that in nature only those states are realised for which the wave function is anti-symmetrical in the coordinates of any two electrons. This would exclude the curve s from consideration if it were not for the spin of the electrons. As explained in Art. 3, the wave function symmetrical in the positional coordinates of the two electrons, which belongs to the curve s, can be supplemented by a function of the spin coordinates in such a way that the wave function as a whole becomes antisymmetrical in all of the coordinates of the two electrons. As shown in Art. 3, this can be done only in one way so that the resultant molecular state is a singlet state with a total spin of the electrons equal to zero. It is in particular a $^1\Sigma$-state since in the two separate atoms the orbital angular momentum of the electrons and hence the component along the internuclear line vanishes, and since this component, being an integral of the motion, remains zero if the nuclei are approached. The wave function anti-symmetrical in the positional coordinates of the electrons, which belongs to the curve a, on the other hand may be supplemented in three different ways by spin functions so as to remain antisym-metrical in all of the coordinates. The corresponding molecular state is hence a triplet state, in particular for the reasons given above a $^3\Sigma$-state. That from two hydrogen atoms in their normal state we get two molecular states, one a $^1\Sigma$-state, the other a $^3\Sigma$-state, is just what we should expect according to the rules of Art. 3. For in this case the orbital angular momenta of the separate atoms are $L_1 = L_2 = 0$ and the spin angular momenta $S_1 = S_2 = \frac{1}{2}$. But now we have learned in addition that stable molecules may actually exist in the $^1\Sigma$-state, while no molecule formation is possible in the $^3\Sigma$-state, owing to the fact that the energy of two hydrogen atoms approaching each other along the curve a of Fig. 16 increases steadily.

Against the procedure just outlined there may be levied the criticism that in the calculation of the energy values as a function of ρ only the first order perturbation effects have been taken into account. For small values of ρ the higher order perturbation terms will naturally become important so that the first order approximation gives very rough results. But for very large values of ρ too it becomes inaccurate. For the first order modification of the energy values on separating the nuclei decreases exponentially with ρ, while the second order effect, as shown by Wang (350), is given by

$$W^{(2)} = -\frac{243}{28}\frac{e^2 a^5}{\rho^6}, \quad a = \frac{h^2}{4\pi^2 m e^2},$$

decreasing hence algebraically. This term represents the Van der Waals interaction between two normal H-atoms, giving rise to an attraction. For normal helium atoms, from which as we shall see below no molecule can be formed, the corresponding term gives rise to the *cohesive forces* which make it possible to liquefy the gas at temperatures sufficiently low. From the above it follows that for very large values of ρ the second order effect is actually more important than the first order effect. Nevertheless the conclusions as to the possibility of molecule formation reached above may be expected to remain valid. For there will be a region of intermediate values of ρ where the first order effect should predominate, and that will mean that one of the two energy values represented in Fig. 15 as functions of ρ will in that region actually come to lie below the energy value of the two separate normal hydrogen atoms, which is all that is required for the possibility of molecule formation.

It is interesting to note that in the wave functions (1) and (2) we have two terms, one corresponding to electron 1 being with nucleus I and electron 2 with nucleus II, the other to electron 2 being with nucleus I and electron 1 with nucleus II. It becomes hence impossible to say that one of the electrons is associated with the first nucleus, the other with the second. *The electrons must rather be conceived of as being shared by the two nuclei*, being in every respect equivalent.

We come next to the discussion of the *behaviour of two helium*

atoms in their normal state. This problem has also been investigated by Heitler and London (338) with the help of the same method
just discussed in the case of hydrogen. Here we have two equal
doubly charged nuclei and four electrons, two with each nucleus to
start with, bound in the state of least energy.

If we call the nuclei I and II and the electrons 1, 2, 3, 4, and if
we first disregard their spin, we have altogether six wave functions
for the system as a whole as long as the internuclear distance is
very great, corresponding to the different ways the electrons may
be distributed on the two nuclei, two with each. They are in a
notation analogous to that previously used,

$$\left. \begin{array}{ll} \Psi_{I12}\ \Psi_{II34}, & \Psi_{I34}\ \Psi_{II12}, \\ \Psi_{I13}\ \Psi_{II42}, & \Psi_{I42}\ \Psi_{II13}, \\ \Psi_{I14}\ \Psi_{II23}, & \Psi_{I23}\ \Psi_{II14}, \end{array} \right\} \quad \dots\dots\dots\dots(3)$$

and belong all to the same energy value as long as the nuclei are
at very great distances apart.

When the nuclei are brought closer together, this energy value
is again split into component levels with each of which there is
associated a definite linear combination of the functions given above,
which has certain symmetry properties as regards the interchange
of the positional coordinates of the electrons. Here too by a
perturbation calculation, considering only first order effects, the
way in which each component energy level depends upon the
internuclear distance ρ may be determined.

If now the spins of the electrons are taken into account, it appears
that only one of the functions composed linearly of the functions
(3) can be supplemented by functions of the spin coordinates in
such a way as to become antisymmetrical in all the coordinates of
any pair of electrons, and that this can be accomplished in only
one way. The resultant molecular level is thus a singlet level, in
particular a $^1\Sigma$-level since the orbital angular momentum of the
electrons in two separate normal helium atoms vanishes. This
conclusion is in harmony with the rules of Art. 3, which tell us

what kind of molecular levels are to be expected from given levels of the separate atoms, as for two helium atoms in their normal state the orbital angular momenta L_1 and L_2 of the electrons as well as their total spins S_1 and S_2 vanish.

The perturbation calculation carried out by Heitler and London shows that the first order effects give for the dependence of the energy value of this $^1\Sigma$-state, the only one which can occur in nature, on the internuclear distance a curve of the same type as curve a in Fig. 16. Such a curve represents a repulsion between two normal helium atoms, and molecule formation is thus impossible for them. There may again be raised the objection that the consideration of the first order perturbation effects is altogether too rough.

From the spectroscopic evidence we know that there exist bands emitted by helium molecules. This result might seem at first at variance with the conclusion arrived at above. The difficulty is removed if it be remembered that our considerations apply only to two helium atoms in their normal states. There is no reason why a normal and an excited helium atom or two excited helium atoms should be incapable of combining to form a molecule, and it is to molecules of this kind that the observed bands must be ascribed.

Comparing the two cases of hydrogen and helium we see that the fundamental difference in their behaviour as regards the possibility of molecule formation is closely connected with the fact of hydrogen atoms having a resultant spin and of helium atoms not having any. Already in the early days of chemical science the chemists introduced the concept of *valency* to describe the faculty of certain atoms to enter into chemical union with other atoms. Figuratively the valency of an atom was conceived of as a number of hooks by which the atom might be attached to other atoms, also carrying hooks. The rare gas atoms according to the chemical evidence were to be considered as having no hooks, monovalent, divalent, ... atoms as having one, two, ... hooks respectively. In the *structure formulae* of chemistry these hooks or links were represented by dashes drawn between the symbols of the elements. From the difference between hydrogen and helium one is naturally led to the hypothesis that the resultant spin of an atom may be the physical reality corresponding to the concept of valency, and that the tendency of the spins of the individual atoms to form a vanishing resultant in the molecule, as we found to be the case in hydrogen, is the underlying cause of what the chemist calls *saturation of valencies* and represents in his model or diagram by the joining of the hooks or the connecting lines between the symbols of the elements. The development of this idea is due to Heitler (339), (340), and London (346), (347), and will be discussed somewhat further in the following article.

28. The General Theory of Homopolar Compounds.

The extension of the results obtained in the last article for two normal hydrogen or helium atoms to include other pairs of atoms and also to embrace the interaction of more than two atoms on the basis of the ideas sketched briefly at the end may be formulated as follows:

As pointed out in Art. 3 the wave functions of the interacting atoms have certain symmetry properties with regard to the interchange of the coordinates of the electrons, and the total spin of the atom depends on the class of symmetry to which the wave function of the stationary state under consideration belongs. From given states of the atoms there results in general a number of energy levels when the nuclei are brought together as described in Art. 3 for the special case of two atoms. For each of these states the spins of the atoms form a resultant spin. For some of them the energy will decrease when the atoms approach, while for others it will increase. In the former case molecules can be formed, in the latter elastic reflection of the atoms will take place.

In the case of two hydrogen atoms in their normal state we have seen that there resulted two energy levels, and that the one in which the spins of the atoms cancelled, the singlet level, led to molecule formation, while the other, for which the spins were parallel, the triplet level, led to elastic reflection. Heitler and London made the assumption that quite generally those states in which a complete cancellation of the spin vectors of the atoms takes place have energy curves lying below those corresponding to states with partial cancellation of the spins or no cancellation at all. Atoms having zero spin should, just as normal helium, be unable to enter into chemical combination.

The hypothetical extension of the results found for hydrogen and helium to other atoms is probably true in a great many cases. Indeed it is found from band spectra that most stable diatomic molecules have zero resultant spin. The rule, however, is by no means rigorous, and definite exceptions are known. Thus the normal state of O_2 is a $^3\Sigma$-state, while a $^1\Sigma$-state lies higher although for it the cancellation of the spins is complete. Other examples of such exceptions are found among the molecular levels arising from a normal hydrogen atom and a hydrogen atom in the state with principal quantum number 2, as shown by Kemble and Zener [345] and Hylleras [344]. Here again states for which the spins cancel completely or for which the valencies are completely saturated are not the ones leading to the firmest binding. Computations of the energy levels of H_2 as a function of ρ have been also made by Condon [337], Wang [351], and Zener and Guillemin [352], and of the energy levels of H_2^+ by Morse and Stueckelberg [348]

BIBLIOGRAPHY

THEORY OF BAND SPECTRA

1. BIRGE and SPONER, *Phys. Rev.* **28**, 259, 1926,
2. BIRGE, *Nature*, **124**, 13, 1929.
3. ———, *ibid.* **124**, 182, 1929.
4. ———, *Phys. Rev.* **34**, 379, 1929.
5. BONHOEFFER and FARKAS, *Zeit. f. phys. Chem.* **134**, 337, 1927.
6. BORN and OPPENHEIMER, *Ann. d. Phys.* **84**, 475, 1927.
7. BRESTER, *Kristallsymmetrie und Reststrahlen*, Diss. Utrecht, 1923.
8. BURRAU, K. *Danske Vid. Selsk., Math.-fys. Medd.* **7**, No. 14, 1927.
9. CONDON, *Phys. Rev.* **28**, 1182, 1926.
10. ———, *Proc. Nat. Ac.* **13**, 462, 1927.
11. ———, *Phys. Rev.* **32**, 858, 1928.
12. DENNISON, *Ast. J.* **62**, 84, 1925.
13. ———, *Phil. Mag.* **1**, 195, 1926.
14. ———, *Phys. Rev.* **28**, 318, 1926.
15. ———, *ibid.* **31**, 503, 1928.
16. DIEKE, *Zeit. f. Phys.* **33**, 161, 1925.
17. ———, *ibid.* **57**, 71, 1929.
18. FOWLER, *Phil. Mag.* **49**, 1272, 1925.
19. FRANCK, *Trans. Faraday Soc.* **21**, 1925.
20. FUES, *Ann. d. Phys.* **80**, 367, 1926.
21. ———, *ibid.* **81**, 218, 1926.
22. GUILLEMIN and ZENER, *Proc. Nat. Ac.* **15**, 314, 1929.
23. HEISENBERG, *Zeit. f. Phys.* **41**, 239, 1927.
24. HEITLER and HERZBERG, *Naturw.* **17**, 673, 1929.
25. HEURLINGER, *Phys. Zeit.* **20**, 188, 1919.
26. HILL and VAN VLECK, *Phys. Rev.* **32**, 250, 1928.
27. HÖNL and LONDON, *Zeit. f. Phys.* **33**, 803, 1925.
28. HULTHÉN, *ibid.* **46**, 349, 1927.
29. HUND, *ibid.* **36**, 657, 1926.
30. ———, *ibid.* **40**, 742, 1927.
31. ———, *ibid.* **42**, 93, 1927.
32. ———, *ibid.* **43**, 788, 1927.
33. ———, *ibid.* **51**, 759, 1928.
34. ITTMANN, *Physica*, **9**, 305, 1929.
35. KEMBLE, *Zeit. f. Phys.* **35**, 286, 1925.
36. ———, *Phys. Rev.* **30**, 387, 1927.
37. KING and BIRGE, *Nature*, **124**, 127, 1929.
38. ——— ———, *Phys. Rev.* **34**, 376, 1929.

39. KRAMERS, *Zeit. f. Phys.* **53**, 422, 1929.
40. —— , *ibid.* **53**, 429, 1929.
41. KRAMERS and ITTMANN, *ibid.* **53**, 553, 1929.
42. —— ——, *ibid.* **58**, 217, 1929.
43. KRATZER, *ibid.* **3**, 289, 1920.
44. —— , *ibid.* **3**, 460, 1920.
45. KRONIG and RABI, *Nature*, **118**, 805, 1926.
46. —— ——, *Phys. Rev.* **29**, 262, 1927.
47. KRONIG, *Zeit. f. Phys.* **46**, 814, 1927.
48. —— , *ibid.* **50**, 347, 1928.
49. —— , *Phys. Rev.* **31**, 195, 1928.
50. —— , *Naturw.* **16**, 335, 1928.
51. LENZ, *Verh. d. deut. phys. Ges.* **31**, 632, 1919.
52. LOOMIS, *Ast. J.* **52**, 248, 1920.
53. MECKE, *Phys. Zeit.* **25**, 597, 1924.
54. —— , *Zeit. f. Phys.* **28**, 261, 1924.
55. —— , *Phys. Zeit.* **26**, 217, 1925.
56. —— , *Zeit. f. Phys.* **31**, 709, 1925.
57. —— , *ibid.* **32**, 823, 1925.
58. —— , *ibid.* **36**, 795, 1926.
59. MECKE and GUILLERY, *Phys. Zeit.* **28**, 514, 1927.
60. MECKE, *Zeit. f. Phys.* **42**, 390, 1927.
61. MENSING, *ibid.* **36**, 814, 1926.
62. MULLIKEN, *Phys. Rev.* **25**, 119, 1925.
63. —— , *ibid.* **26**, 561, 1925.
64. —— , *ibid.* **28**, 481, 1926.
65. —— , *ibid.* **28**, 1202, 1926.
66. —— , *Proc. Nat. Ac.* **12**, 144, 1926.
67. —— , *ibid.* **12**, 151, 1926.
68. —— , *ibid.* **12**, 338, 1926.
69. —— , *Phys. Rev.* **29**, 391, 1927.
70. —— , *ibid.* **29**, 637, 1927.
71. —— , *ibid.* **30**, 138, 1927.
72. —— , *ibid.* **30**, 785, 1927.
73. —— , *ibid.* **32**, 186, 1928.
74. —— , *ibid.* **32**, 388, 1928.
75. —— , *ibid.* **32**, 761, 1928.
76. —— , *ibid.* **33**, 507, 1929.
77. —— , *ibid.* **33**, 730, 1929.
78. OPPENHEIMER, *Proc. Camb. Phil. Soc.* **23**, 327, 1926.
79. RADEMACHER and REICHE, *Zeit. f. Phys.* **41**, 453, 1927
80. REICHE, *ibid.* **39**, 444, 1926.
81. SCHWARZSCHILD, *Sitz. preuss. Ak.* 1916, 548.

82. SLATER, *Proc. Nat. Ac.* **13**, 423, 1927.
83. VAN VLECK, *Phys. Rev.* **33**, 467, 1929.
84. WANG, *ibid.* **34**, 243, 1929.
85. WIGNER and WITMER, *Zeit. f. Phys.* **51**, 859, 1928.

ANALYSIS OF BAND SPECTRA.
ELECTRONIC BANDS

H_2 86. ALLEN, *Proc. Roy. Soc.* **A 106**, 69, 1924.
87. ALLEN and SANDEMAN, *ibid.* **A 114**, 293, 1926.
88. —— ——, *ibid.* **A 116**, 312, 1927.
89. BIRGE, *Proc. Nat. Ac.* **14**, 12, 1928.
90. CONDON and SMYTHE, *ibid.* **14**, 871, 1928.
91. DIEKE, *Phil. Mag.* **1**, 173, 1925.
92. ——, *Zeit. f. Phys.* **32**, 180, 1925.
93. DIEKE and HOPFIELD, *ibid.* **40**, 299, 1927.
94. —— ——, *Phys. Rev.* **30**, 400, 1927.
95. DIEKE, *Zeit. f. Phys.* **55**, 447, 1929.
96. FINKELNBURG and MECKE, *ibid.* **54**, 198, 1929.
97. —— ——, *ibid.* **54**, 597, 1929.
98. HORI, *ibid.* **44**, 834, 1927.
99. KAPUSCINSKI and EYMERS, *Proc. Roy. Soc.* **A 122**, 58, 1929.
100. KEMBLE and GUILLEMIN, *Proc. Nat. Ac.* **14**, 782, 1928.
101. RICHARDSON and TANAKA, *Proc. Roy. Soc.* **A 106**, 643, 1924.
102. —— ——, *ibid.* **A 107**, 602, 1925.
103. RICHARDSON, *ibid.* **A 111**, 714, 1926.
104. —— , *ibid.* **A 113**, 368, 1926.
105. —— , *ibid.* **A 114**, 643, 1926.
106. —— , *ibid.* **A 115**, 528, 1927.
107. —— , *ibid.* **A 116**, 484, 1927.
108. RICHARDSON and DAS, *ibid.* **A 122**, 688, 1929.
109. RICHARDSON and DAVIDSON, *ibid.* **A 123**, 54, 1929.
110. —— ——, *ibid.* **A 123**, 466, 1929.
111. —— ——, *ibid.* **A 124**, 50, 1929.
112. —— ——, *ibid.* **A 124**, 69, 1929.
113. —— ——, *ibid.* **A 125**, 23, 1929.
114. RICHARDSON and DAS, *ibid.* **A 125**, 309, 1929.
115. SANDEMAN, *ibid.* **A 108**, 607, 1925.
116. —— , *ibid.* **A 110**, 326, 1926.
117. SCHAAFSMA and DIEKE, *Zeit. f. Phys.* **55**, 164, 1929.
118. WEIZEL, *ibid.* **52**, 175, 1928.
119. —— , *ibid.* **55**, 483, 1929.
120. WERNER, *Proc. Roy. Soc.* **A 113**, 107, 1926.

121. WINANS and STUECKELBERG, *Proc. Nat. Ac.* **14**, 867, 1928.
122. WITMER, *ibid.* **12**, 238, 1926.
123. ——, *Phys. Rev.* **28**, 1223, 1926.
He₂ 124. CURTIS, *Proc. Roy. Soc.* **A 101**, 38, 1923.
125. ——, *ibid.* **A 103**, 315, 1923.
126. CURTIS and LONG, *ibid.* **A 108**, 513, 1925.
127. CURTIS, *ibid.* **A 118**, 157, 1928.
128. CURTIS and HARVEY, *ibid.* **A 121**, 381, 1928.
129. —— ——, *ibid.* **A 125**, 484, 1929.
130. DIEKE, TAKAMINE and SUGA, *Zeit. f. Phys.* **49**, 637, 1928.
131. DIEKE, IMANISHI and TAKAMINE, *ibid.* **54**, 826, 1929.
132. —— —— ——, *ibid.* **57**, 305, 1929.
133. DIEKE, *Nature*, **123**, 446, 1929.
134. KRATZER, *Zeit. f. Phys.* **16**, 353, 1923.
135. MULLIKEN, *Proc. Nat. Ac.* **12**, 158, 1926.
136. WEIZEL and FÜCHTBAUER, *Zeit. f. Phys.* **44**, 431, 1927.
137. WEIZEL, *ibid.* **51**, 328, 1928.
138. ——, *ibid.* **54**, 321, 1929.
139. WEIZEL and PESTEL, *ibid.* **56**, 197, 1929.
O₂ 140. SHEA, *Phys. Rev.* **30**, 825, 1927.
N₂ 141. BIRGE and HOPFIELD, *Ast. J.* **68**, 257, 1928.
142. LINDAU, *Zeit. f. Phys.* **26**, 343, 1924.
143. SPONER, *Proc. Nat. Ac.* **13**, 100, 1927.
144. ——, *Zeit. f. Phys.* **41**, 611, 1927.
N₂⁺ 145. FASSBENDER, *Zeit. f. Phys.* **30**, 73, 1924.
146. HERZBERG, *ibid.* **49**, 761, 1928.
147. ——, *Ann. d. Phys.* **86**, 189, 1928.
148. MERTON and PILLEY, *Phil. Mag.* **50**, 195, 1925.
149. ORNSTEIN and VAN WIJK, *Zeit. f. Phys.* **49**, 315, 1928.
O₂ 150. DIEKE and BABCOCK, *Proc. Nat. Ac.* **13**, 670, 1927.
151. ELLSWORTH and HOPFIELD, *Phys. Rev.* **29**, 79, 1927.
152. LEIFSON, *Ast. J.* **63**, 73, 1926.
153. MULLIKEN, *Phys. Rev.* **32**, 880, 1928.
154. OSSENBRÜGGEN, *Zeit. f. Phys.* **49**, 167, 1928.
155. RASETTI, *Proc. Nat. Ac.* **15**, 411, 1929.
156. RUNGE, *Physica*, **1**, 254, 1921.
O₂⁺ 157. FRERICHS, *Zeit. f. Phys.* **35**, 683, 1926.
F₂ 158. GALE and MONK, *Ast. J.* **69**, 77, 1929.
Na₂ 159. FREDRICKSON and WATSON, *Phys. Rev.* **30**, 429, 1927.
160. FREDRICKSON, *ibid.* **34**, 207, 1929.
161. LOOMIS, *ibid.* **31**, 323, 1928.
162. LOOMIS and WOOD, *ibid.* **32**, 223, 1928.
163. LOOMIS and NILE, *ibid.* **32**, 873, 1928.

S_2 164. HENRI and TEVES, *Nature*, **114**, 894, 1924.

165. HENRI and WURMSER, *J. de Phys.* **8**, 289, 1927.

166. ROSEN, *Zeit. f. Phys.* **43**, 69, 1927.

167. —— , *ibid.* **48**, 545, 1928.

Cl_2 168. ELLIOTT, *Proc. Roy. Soc.* **A 123**, 629, 1929.

169. KUHN, *Zeit. f. Phys.* **39**, 77, 1926.

Br_2 170. KUHN, *Zeit. f. Phys.* **39**, 77, 1926.

I_2 171. KEMBLE and WITMER, *Phys. Rev.* **28**, 633, 1926.

172. KRATZER and SUDHOLT, *Zeit. f. Phys.* **33**, 144, 1925.

173. LOOMIS, *Phys. Rev.* **29**, 112, 1927.

174. MECKE, *Ann. d. Phys.* **71**, 104, 1923.

175. PRINGSHEIM and ROSEN, *Zeit. f. Phys.* **50**, 1, 1928.

176. WOOD and LOOMIS, *Phil. Mag.* **6**, 231, 1928.

LiH 177. WATSON, *Phys. Rev.* **34**, 372, 1929.

BeH 178. HULTHÉN, *Ark. f. Math., Astr. och Fys.* **21**, 1929.

179. WATSON, *Phys. Rev.* **32**, 600, 1928.

BeH+ 180. WATSON, *Phys. Rev.* **32**, 600, 1928.

181. —— , *ibid.* **34**, 372, 1929.

CH 182. KRATZER, *Zeit. f. Phys.* **23**, 298, 1925.

OH 183. JACK, *Proc. Roy. Soc.* **A 115**, 373, 1927.

184. ——, *ibid.* **A 118**, 647, 1928.

185. ——, *ibid.* **A 120**, 222, 1928.

186. WATSON, *Ast. J.* **60**, 145, 1924.

187. —— , *Nature*, **117**, 157, 1926.

MgH 188. HULTHÉN, *Ark. f. Math., Astr. och Fys.* **21**, 1929.

189. PEARSE, *Proc. Roy. Soc.* **A 122**, 442, 1929.

190. WATSON and RUDNICK, *Ast. J.* **63**, 20, 1926.

191. —— ——, *Phys. Rev.* **29**, 413, 1927.

192. WATSON, *ibid.* **34**, 372, 1929.

MgH+ 193. PEARSE, *Proc. Roy. Soc.* **A 125**, 157, 1929.

AlH 194. BENGTSSON, *Zeit. f. Phys.* **51**, 889, 1928.

195. BENGTSSON and HULTHÉN, *ibid.* **52**, 275, 1928.

196. ERIKSSON and HULTHÉN, *ibid.* **34**, 775, 1925.

CaH 197. HULTHÉN, *Phys. Rev.* **29**, 97, 1927.

198. —— , *Ark. f. Math., Astr. och Fys.* **21**, 1929.

199. MULLIKEN, *Phys. Rev.* **25**, 509, 1925.

CuH 200. BENGTSSON, *Zeit. f. Phys.* **20**, 229, 1923.

201. FRERICHS, *ibid.* **20**, 170, 1923.

202. HULTHÉN and ZUMSTEIN, *Phys. Rev.* **28**, 13, 1926.

ZnH 203. HULTHÉN, *Ark. f. Math., Astr. och Fys.* **21**, 1929.

AgH 204. BENGTSSON, *Ark. f. Math., Astr. och Fys.* **18**, 1925.

205. HULTHÉN and ZUMSTEIN, *Phys. Rev.* **28**, 13, 1926.

CdH 206. HULTHÉN, *Ark. f. Math., Astr. och Fys.* **21**, 1929.

AuH 207. BENGTSSON, *Ark. f. Math., Astr. och Fys.* **18**, 1925.

208. HULTHÉN and ZUMSTEIN, *Phys. Rev.* **28**, 13, 1926.

HgH 209. HULTHÉN, *Zeit. f. Phys.* **32**, 32, 1925.

210. ——, *ibid.* **50**, 319, 1928.

211. ——, *Ark. f. Math., Astr. och Fys.* **21**, 1929.

212. KAPUSCINSKI and EYMERS, *Zeit. f. Phys.* **54**, 246, 1929.

213. LUDLOFF, *ibid.* **34**, 485, 1925.

CN 214. BIRGE, *Phys. Rev.* **11**, 136, 1918.

215. DIEKE, *Physica,* **5**, 178, 1925.

216. JEVONS, *Proc. Roy. Soc.* **A 112**, 407, 1926.

217. JENKINS, *Phys. Rev.* **31**, 539, 1928.

218. ROSENTHAL and JENKINS, *Proc. Nat. Ac.* **15**, 381, 1929.

SiN 219. JENKINS and LASZLO, *Proc. Roy. Soc.* **A 122**, 103, 1929

220. MULLIKEN, *Phys. Rev.* **26**, 319, 1926.

BeO 221. BENGTSSON, *Ark. f. Math., Astr. och Fys.* **20**, 1928.

222. ROSENTHAL and JENKINS, *Phys. Rev.* **33**, 163, 1929.

BO 223. JENKINS, *Proc. Nat. Ac.* **13**, 496, 1927.

224. MULLIKEN, *Phys. Rev.* **25**, 259, 1925.

CO 225. ASUNDI, *Proc. Roy. Soc.* **A 124**, 277, 1929.

226. BIRGE, *Phys. Rev.* **28**, 1157, 1926.

227. DUFFENDACK and FOX, *Ast. J.* **65**, 220, 1927.

228. HULTHÉN, *Ann. d. Phys.* **71**, 43, 1923.

229. HERZBERG, *Zeit. f. Phys.* **52**, 815, 1929.

230. JOHNSON and ASUNDI, *Proc. Roy. Soc.* **A 123**, 560, 1929.

231. LEIFSON, *Ast. J.* **63**, 73, 1926.

CO+ 232. BIRGE, *Phys. Rev.* **28**, 1157, 1926.

233. DAWSON and KAPLAN, *ibid.* **34**, 379, 1929.

NO 234. GUILLERY, *Zeit. f. Phys.* **42**, 121, 1927.

235. JENKINS, BARTON and MULLIKEN, *Phys. Rev.* **30**, 150, 1927.

AlO 236. POMEROY, *Phys. Rev.* **29**, 59, 1927.

SO 237. HENRI and WOLFF, *J. de Phys.* **10**, 81, 1929.

TiO 238. CHRISTEY, *Phys. Rev.* **33**, 701, 1929.

CS 239. JEVONS, *Proc. Roy. Soc.* **A 117**, 351, 1928.

BeF 240. DATTA, *Proc. Roy. Soc.* **A 101**, 187, 1922.

CuF 241. RITSCHL, *Zeit. f. Phys.* **42**, 172, 1927.

CuCl 242. RITSCHL, *Zeit. f. Phys.* **42**, 172, 1927.

AgCl 243. FRANCK and KUHN, *Zeit. f. Phys.* **44**, 607, 1927.

SnCl 244. FERGUSON, *Phys. Rev.* **32**, 607, 1928.

245. JEVONS, *Proc. Roy. Soc.* **A 110**, 365, 1926.

AuCl 246. FERGUSON, *Phys. Rev.* **31**, 969, 1928.

ICl 247. GIBSON and RAMSPERGER, *Phys. Rev.* **30**, 598, 1927.

248. GIBSON, *Zeit. f. Phys.* **50**, 692, 1928.

249. WILSON, *Phys. Rev.* **32**, 611, 1928.

CuBr 250. RITSCHL, *Zeit. f. Phys.* **42**, 172, 1927.

AgBr 251. FRANCK and KUHN, *Zeit. f. Phys.* **44**, 607, 1927.

CuI 252. MULLIKEN, *Phys. Rev.* **26**, 1, 1925.

253. RITSCHL, *Zeit. f. Phys.* **42**, 172, 1927.

AgI 254. FRANCK and KUHN, *Zeit. f. Phys.* **43**, 164, 1927.

Stark Effect 255. MACDONALD, *Proc. Roy. Soc.* **A 123**, 103, 1929.

Zeeman Effect 256. CRAWFORD, *Phys. Rev.* **33**, 341, 1929.

257. CURTIS and JEVONS, *Proc. Roy. Soc.* **A 120**, 110, 1928.

258. KEMBLE, MULLIKEN and CRAWFORD, *Phys. Rev.* **30**, 438, 1927.

259. WATSON and PERKINS, *ibid.* **30**, 592, 1927.

VIBRATIONAL BANDS

CO 260. SNOW and RIDEAL, *Proc. Roy. Soc.* **A 125**, 462, 1929.

NO 261. SNOW, RAWLINS and RIDEAL, *Proc. Roy. Soc.* **A 124**, 453, 1929.

HF 262. IMES, *Ast. J.* **50**, 251, 1919.

HCl 263. BOURGIN, *Phys. Rev.* **29**, 794, 1927.

264. —— , *ibid.* **32**, 237, 1928.

265. BRINSMADE and KEMBLE, *Proc. Nat. Ac.* **3**, 420, 1917.

266. COLBY and MEYER, *Ast. J.* **53**, 300, 1921.

267. COLBY, MEYER and BRONK, *Ast. J.* **57**, 7, 1923.

268. DUNHAM, *Phys. Rev.* **34**, 438, 1929.

269. MEYER and LEVIN, *ibid.* **34**, 44, 1929.

HBr 270. IMES, *Ast. J.* **50**, 251, 1919.

NH₃ 271. BARKER, *Phys. Rev.* **33**, 684, 1929.

272. HETTNER, *Zeit. f. Phys.* **31**, 273, 1925.

273. ROBERTSON and FOX, *Proc. Roy. Soc.* **A 120**, 128, 1928.

PH₃ 274. ROBERTSON and FOX, *Proc. Roy. Soc.* **A 120**, 128, 1928.

AsH₃ 275. ROBERTSON and FOX, *Proc. Roy. Soc.* **A 120**, 128, 1928.

CH₄ 276. COOLEY, *Ast. J.* **63**, 73, 1925.

ROTATIONAL BANDS

HF 277. CZERNY, *Zeit. f. Phys.* **44**, 235, 1927.

HCl 278. BADGER, *Proc. Nat. Ac.* **13**, 408, 1927.

279. CZERNY, *Zeit. f. Phys.* **34**, 227, 1925.

280. —— , *ibid.* **44**, 235, 1927.

HBr 281. CZERNY, *Zeit. f. Phys.* **44**, 235, 1927.

HI 282. CZERNY, *Zeit. f. Phys.* **44**, 235, 1927.

NH₃ 283. BADGER and CARTWRIGHT, *Phys. Rev.* **33**, 692, 1929.

OPTICAL PROPERTIES OF MOLECULAR GASES

284. BORN, *Naturw.* **16**, 673, 1928.
285. DARWIN and WATSON, *Proc. Roy. Soc.* **A 114**, 474, 1927.
286.⸲ DEBYE, BEWILOGUA and EHRHARDT, *Phys. Zeit.* **30**, 84, 1929.
287. DIEKE, *Nature*, **123**, 564, 1929.
288. DICKINSON, DILLON and RASETTI, *Phys. Rev.* **34**, 582, 1929.
289. EHRENFEST, *Proc. Amst. Ac.* **17**, 1184, 1915.
290. HERZFELD, *Ann. d. Phys.* **69**, 369, 1922.
291. HILL and KEMBLE, *Proc. Nat. Ac.* **15**, 387, 1929.
292. LANGER, *Nature*, **123**, 345, 1929.
293. MANNEBACK, *Naturw.* **17**, 364, 1929.
294. RASETTI, *Nature*, **123**, 205, 1929.
295. ——— , *ibid.* **123**, 757, 1929.
296. ——— , *ibid.* **124**, 93, 1929.
297. ——— , *Proc. Nat. Ac.* **15**, 234, 1929.
298. ——— , *ibid.* **15**, 515, 1929.
299. ——— , *Phys. Rev.* **34**, 367, 1929.
300. SMEKAL, *Naturw.* **16**, 612, 1928.
301. SZIVESSY, *Zeit. f. Phys.* **26**, 323, 1924.
302. VAN VLECK, *Proc. Nat. Ac.* **15**, 754, 1929.
303. WOOD, *Nature*, **123**, 279, 1929.
304. ——— , *Phil. Mag.* **7**, 744, 1929.

DIELECTRIC AND MAGNETIC PROPERTIES OF MOLECULAR GASES

305. BAUER and PICCARD, *J. de Phys.* **1**, 97, 1920.
306. BITTER, *Proc. Nat. Ac.* **15**, 638, 1929.
307. BLÜH, *Phys. Zeit.* **27**, 226, 1926.
308. DEBYE, *Phys. Zeit.* **13**, 97, 1912.
309. DEBYE and MANNEBACK, *Nature*, **119**, 83, 1927.
310. HØJENDAHL, *Diss. Copenhagen*, 1928.
311. KRONIG, *Proc. Nat. Ac.* **12**, 488, 1926.
312. ——— , *ibid.* **12**, 608, 1926.
313. LANGEVIN, *J. de Phys.* **4**, 678, 1905.
314. MANNEBACK, *Phys. Zeit.* **27**, 563, 1926.
315. ——— , *ibid.* **28**, 72, 1927.
316. MENSING and PAULI, *ibid.* **27**, 509, 1926.
317. NIESSEN, *Phys. Rev.* **34**, 253, 1929.
318. SONÉ, *Tohoku Univ. Sci. Rep.* **11**, 139, 1922.
319. TAYLOR, *J. Amer. Chem. Soc.* **48**, 854, 1926.

320. VAN VLECK, *Nature*, **118**, 226, 1926.
321. —— , *Phys. Rev.* **29**, 727, 1927.
322. —— , *ibid.* **30**, 31, 1927.
323. —— , *Nature*, **119**, 670, 1927.
324. —— , *Phys. Rev.* **31**, 587, 1928.
325. VAN VLECK and FRANK, *Proc. Nat. Ac.* **15**, 539, 1929.
326. WANG, *ibid.* **13**, 798, 1927.
327. WILLS and HECTOR, *Phys. Rev.* **23**, 209, 1924.
328. WREDE, *Zeit. f. Phys.* **44**, 261, 1927.

THERMAL PROPERTIES OF
MOLECULAR GASES

329. BONHOEFFER and HARTECK, *Naturw.* **17**, 182, 1929.
330. —— ——, *Sitz. preuss. Ak.* 1929, 103.
331. —— ——, *Zeit. f. phys. Chem.* **B 4**, 113, 1929.
332. DENNISON, *Proc. Roy. Soc.* **A 115**, 483, 1927.
333. EUCKEN, *Naturw.* **17**, 182, 1929.
334. McCREA, *Proc. Camb. Phil. Soc.* **24**, 80, 1928.
335. MULHOLLAND, *ibid.* **24**, 280, 1928.
336. PARTINGTON and SHILLING, *The specific heats of gases*, Benn, 1924.

MOLECULE FORMATION

337. CONDON, *Proc. Nat. Ac.* **13**, 466, 1927.
338. HEITLER and LONDON, *Zeit. f. Phys.* **44**, 455, 1927.
339. HEITLER, *ibid.* **47**, 835, 1928.
340. —— , *ibid.* **51**, 804, 1928.
341. HEITLER and HERZBERG, *ibid.* **53**, 52, 1929.
342. HUND, *ibid.* **31**, 81, 1925.
343. —— , *ibid.* **32**, 1, 1925.
344. HYLLERAS, *ibid.* **51**, 150, 1928.
345. KEMBLE and ZENER, *Phys. Rev.* **33**, 512, 1929.
346. LONDON, *Zeit. f. Phys.* **46**, 455, 1928.
347. —— , *ibid.* **50**, 24, 1928.
348. MORSE and STUECKELBERG, *Phys. Rev.* **33**, 932, 1929.
349. SUGIURA, *Zeit. f. Phys.* **45**, 484, 1927.
350. WANG, *Phys. Zeit.* **28**, 663, 1927.
351. —— , *Phys. Rev.* **31**, 579, 1928.
352. ZENER and GUILLEMIN, *Phys. Rev.* **34**, 999, 1929.

SUBJECT INDEX

(Abbreviations: d. = diatomic, p. = polyatomic, m. = molecule)

Alternation of intensities, 89, 94 *et seq.*
Antisymmetrical levels of d.m., 68 *et seq.*
Atom, 2

Band head, 88
Band spectra, 2
Band system, 85

Chemical binding of heteropolar m., 141 *et seq.*; of homopolar m., 144 *et seq.*
Coherent scattering of gases, 103 *et seq.*, 116
Cohesive forces in gases, 148
Combination principle, 3
Coordinates of d.m., 6 *et seq.*
Coordination of molecular to atomic levels, 23 *et seq.*

Diatomic m., 2
Dielectric constant of gases, 125 *et seq.*
Dielectric polarisation of gases, 121 *et seq.*
Dispersion of gases, 113 *et seq.*
Dissociation of d.m., 28

Electron, 1
Electronic angular momentum, orbital, 18 *et seq.*; spin, 19 *et seq.*
Electronic bands, frequencies, 85 *et seq.*; intensities, 88 *et seq.*; selection rules, 72 *et seq.*; transition probabilities, 78 *et seq.*
Electronic levels of d.m., 5, 17 *et seq.*; inverted, 19, 47 *et seq.*; normal, 19, 47 *et seq.*; of p.m., 38
Electronic quantum numbers of d.m., 5, 17 *et seq.*
Even levels of d.m., 60 *et seq.*

Faraday effect of gases, 117 *et seq.*
Fine structure in d.m., 49 *et seq.*
Frequencies of electronic bands, 85 *et seq.*; of rotational bands, 93; of vibrational bands, 91 *et seq.*
Frequency condition, 3

Hamiltonian function of d.m., 8 *et seq.*

Hamiltonian operator of d.m., 8 *et seq.*
Heteronuclear m., 35
Heteropolar m., 141
Homonuclear m., 35
Homopolar m., 141
Hydrogen molecule ion, 17

Incoherent scattering of gases, 103 *et seq.*
Intensities in electronic bands, 88 *et seq.*; in rotational bands, 93; in Stark and Zeeman effects, 100; in vibrational bands, 92
Inverted electronic levels, 19, 47 *et seq.*
Isotope effect, 28 *et seq.*, 32 *et seq.*, 89
Isotopes, 1, 97

Kerr constant, 118
Kerr effect of gases, 117 *et seq.*

Λ-doubling, 49 *et seq.*
Linear Stark effect of d.m., 36
Linear Zeeman effect of d.m., 37
Line spectra, 2

Magnetic susceptibility of gases, 129 *et seq.*
Magnetisation of gases, 128 *et seq.*
Molecule, 2; diatomic, 2; heteronuclear, 35; heteropolar, 141; homonuclear, 35; homopolar, 141; polyatomic, 2
Multiplicity, 19

Normal electronic levels, 19, 47 *et seq.*
Nuclear spin, 94 *et seq.*
Nucleus, 1

Odd levels of d.m., 60 *et seq.*

Paschen-Back effect of d.m., 37
Perturbations, 53 *et seq.*
Perturbation theory, 40 *et seq.*
Polarisation, dielectric, 121 *et seq.*; of scattered radiation, 105
Polyatomic molecule, 2
Predissociation, 55 *et seq.*
Probability of absorption, 70 *et seq.*

Printed by Printforce, United Kingdom